Ecology and Historical Materialism

This book challenges the widely held view that Marxism is unable to deal adequately with environmental problems. Jonathan Hughes considers the nature of environmental problems and the evaluative perspectives that may be brought to bear on them. He examines Marx's critique of Malthus, his method, and his materialism, interpreting the latter as a recognition of human dependence on nature. Central to the book's argument is an interpretation of the 'development of the productive forces' which takes account of the differing ecological impacts of different productive technologies while remaining consistent with the normative and explanatory roles that this concept plays within Marx's theory. Turning finally to Marx's vision of a society founded on the communist principle 'to each according to his needs', the author concludes that the underlying notion of human need is one whose satisfaction presupposes only a modest and eco-logically feasible expansion of productive output.

JONATHAN HUGHES is Research Fellow in Philosophy and Politics at the University of Manchester.

Studies in Marxism and Social Theory

Edited by G. A. COHEN, JOHN ROEMER AND ERIK OLIN WRIGHT

The series is jointly published by the Cambridge University Press and the Editions de la Maison des Sciences de l'Homme, as part of the joint publishing agreement established in 1977 between the Fondation de la Maison des Sciences de l'Homme and the Syndics of the Cambridge University Press.

The books in the series are intended to exemplify a new paradigm in the study of Marxist social theory. They will not be dogmatic or purely exegetical in approach. Rather, they will examine and develop the theory pioneered by Marx, in the light of the intervening history, and with the tools of non-Marxist social science and philosophy. It is hoped that Marxist thought will thereby be freed from the increasingly discredited methods and presuppositions which are still widely regarded as essential to it, and that what is true and important in Marxism will be more firmly established.

Also in the series

JON ELSTER   *Making Sense of Marx*
ADAM PRZEWORSKI   *Capitalism and Social Democracy*
JOHN ROEMER (ed.)   *Analytical Marxism*
JON ELSTER AND KARL MOENE (eds.)   *Alternatives to Capitalism*
MICHAEL TAYLOR (ed.)   *Rationality and Revolution*
DONALD L. DONMAN   *History, Power, and Ideology*
DAVID SCHWEICKART   *Against Capitalism*
PHILIPPE VAN PARIJS   *Marxism Recycled*
JOHN TORRANCE   *Karl Marx's Theory of Ideas*
G. A. COHEN   *Self-ownership, Freedom, and Equality*
ERIK OLIN WRIGHT   *Class Counts*

Ecology and Historical Materialism

Jonathan Hughes

PUBLISHED BY THE PRESS SYNDICATE OF THE UNIVERSITY OF CAMBRIDGE
The Pitt Building, Trumpington Street, Cambridge, United Kingdom

CAMBRIDGE UNIVERSITY PRESS
The Edinburgh Building, Cambridge CB2 2RU, UK   http://www.cup.cam.ac.uk
40 West 20th Street, New York, NY 10011–4211, USA   http://www.cup.org
10 Stamford Road, Oakleigh, Melbourne 3166, Australia
Ruiz de Alarcón 13, 28014 Madrid, Spain
and Editions de la Maison des Sciences de l'Homme
54 Boulevard Raspail, 75270 Paris Cedex 06

First published 2000

Printed in the United Kingdom at the University Press, Cambridge

Typeset in Palatino 10/13 pt in QuarkXPress® [SE]

A catalogue record for this book is available from the British Library

Library of Congress Cataloguing in Publication data

Hughes, Jonathan R. T.
Ecology and historical materialism / Jonathan Hughes.
    p.   cm.
Includes bibliographical references (p.   ).
ISBN 0 521 66090 4 – ISBN 0 521 66789 5 (pb)
1. Ecology. 2. Historical materialism. I. Title.
QH541.H83 2000
577'.01 21 – dc21   99-040259

ISBN 0 521 66090 4 hardback
ISBN 0 521 66789 5 paperback

ISBN 2 7351 0868 6 hardback (France only)
ISBN 2 7351 0869 4 paperback (France only)

For Katie and Laura

Contents

Acknowledgements

This book has evolved out of a doctoral thesis supervised by Norman Geras and Hillel Steiner, and examined by Ted Benton and Mike Evans. I am grateful to all of them and to the anonymous referees who have commented on all or part of its contents. Particular thanks are due to Andrew Chitty whose extensive and penetrating comments on the penultimate draft have resulted in numerous improvements as well as a keener awareness of the faults that remain. Audiences at two meetings of the PSA Marxism Specialist Group, seminars at Manchester and Wolverhampton, and a conference of the Manchester Centre for Political Thought have also provided helpful suggestions on parts of the argument. An earlier version of chapter 5 appeared in *Studies in Marxism* 2 (1995). My loudest thanks go to Helen Best, who encouraged me to begin this project, and whose continued encouragement, support and patience have been vital to its completion. The book is dedicated to our daughters, Katie and Laura.

Introduction

This book attempts to defend some of the central theses of Marxism – those that make up the theory of historical materialism – against criticisms that have been levelled at it by environmentalists, and to show that historical materialism, suitably interpreted, can provide an explanatory and normative framework for thinking about and developing political responses to the environmental problems that afflict and threaten contemporary societies. The book may therefore be said to involve a confrontation between Marxism and environmentalism, but like any brief summary this formulation is in need of qualification.

To start with, it is an oversimplification to speak of Marxism as a single theory. It is hardly necessary to comment on the diverse range of interpretations that have been applied to the writings of Marx and his collaborator Engels. The nature of that collaboration has also been the source of much dispute, leading to denials of a unitary standpoint in their joint corpus. And within Marx's own works divisions have been discerned between his earlier and later works. The question therefore arises of which version of Marxism it is that is to be placed in confrontation with environmentalism, and consequently it will be one of the main tasks of this book to consider what interpretations are available and which emerge most favourably from that confrontation.

Similarly, it may be observed that environmentalism has many different strands. The terms 'deep' and 'shallow', deriving from Naess's 1973 article, are often used as a convenient way of drawing a distinction between more and less radical strands, but, as John Barry has recently argued, this dichotomy masks a more complex picture.[1] The deep/shallow categorisation is most often defined in evaluative terms, the title of Deep Ecology being

[1] See Barry 1994.

1

claimed by those who ascribe intrinsic value to non-human nature, while those who ascribe to it only instrumental value are labelled – usually by others – shallow. But as we shall see, this terminology is also used in ways which owe more to metaphysical than to evaluative considerations; and, as Barry points out, strands of environmental thought may also be categorised according to their differing economic, institutional and political perspectives, which can cut across the deep/shallow divide. I am less concerned, however, about these differences within environmental thought than I am about the parallel differences within Marxism; my aim is not to choose between the various environmental theorists whose writings will be considered but to use their arguments as a way of raising and developing the challenges that environmental problems pose for Marxism.

The confrontation, then, is asymmetrical. My primary aim is to investigate what the existence of environmental problems means for Marxism and what, if anything, Marxism can contribute to the study and resolution of those problems, and I draw upon a range of environmental literature as a means to this end. Expressed in these terms, and in the face of the widely advertised 'death of Marxism', this project may appear perverse. 'Why Marxism?' it might well be asked. Why not 'Ecology and . . .' liberalism, or communitarianism, or post-modernism even? One answer might be: just because its death is so widely proclaimed. This would not be mere perversity; it is important that intellectual fashions be contested and not simply followed, and that ideas currently out of favour not be forgotten or ignored. Even if, as seems likely following the collapse of political systems and movements supposedly based on his theories, Marx is not in future read as religiously as he has been in the past, it remains important that his ideas not be dismissed but continue to be studied, in order that proper debate about their strengths and weaknesses and their place in the canon of political thought can take place.

This rationale is important but not sufficient, since what I am engaged in is not just a study in the history of ideas but an attempt to relate Marxism to a particular set of contemporary problems. Clearly I am assuming that there are at least *prima facie* grounds for regarding Marxism not only as worthy of study generally, but as a useful framework for the investigation of ecological problems. To see why, consider the fact that, like Marxism, environmentalism is said by some to have had its day. The scare stories of the 1970s and 80s have proved false, and whatever legitimate concerns the environmentalists had have been incorporated into the political mainstream. Or so it is said. But while it is true that expressions of concern about environmental problems have permeated mainstream politics, sapping support for green parties, and that the focus of environmental activism has shifted from political parties to pressure groups which are themselves seen as increasingly integrated into the establishment, there remains a widespread feeling that not enough is being done, and that environmental issues need to become more central to policy-making as the green parties have urged. The feeling that government and corporate expressions of environmental concern are, if not mere window-dressing, then at best peripheral tinkering, subordinated to established political or economic commitments, is evidenced by the willingness of many people to trust the scientific opinions of pressure groups rather than the experts employed by government or business, and by the emergence of new, more confrontational campaigning groups.

One reason, then, for investigating Marxism in relation to ecological problems, is that it may help us to diagnose the weaknesses of green politics and the inadequacy of mainstream responses to ecological problems. Marxists have, for example, criticised the attempts of many ecologists to transcend class divisions and to appeal equally to all humanity. Of course, the existence of ecological problems *is* potentially a threat to everybody, but not to the same extent or with the same degree of immediacy; money can, to a certain extent, buy protection or an escape route.[2] Relatedly, much of green political discourse may, in standard Marxist parlance, be termed utopian for its promulgation of models of a better society constructed and promoted without sufficient attention to the mechanism and agency that are to bring it into existence. And, for Marxism, these same structures of interests that can explain the weakness of ecologism can also explain the inadequacy of the mainstream responses. So perhaps the critical perspective of Marxism can restore the radical edge of ecological politics. It will be argued, however, by environmentalists and others that Marxism is unsuited to this task. This is argued on empirical grounds, by reference to the environmental record of former socialist countries, and it is also argued on theoretical grounds. It is argued that whatever the strengths of a Marxist critique of ecological politics and of the treatment of ecological issues within mainstream politics, Marxist theory is poorly placed to offer an

---

[2] See, for example, Hall 1972b. See also Marx's comment on Utopian Socialism quoted in chapter 1 below, text to note 26. Some greens (e.g. Porritt 1985, p. 116) do acknowledge the differential impact of environmental problems upon different classes, but typically these class differences are downplayed, and do not translate into thoughts about agency or strategy. As Marx writes of the Utopian Socialists: 'They are conscious of caring chiefly for the interests of the working class, as being the most suffering class. Only from the point of view of being the most suffering class does the proletariat exist for them.' (*Manifesto of the Communist Party*, p. 60.)

alternative, since the theory itself has implications – notably those arising from Marx's vision of an abundant future and his commitment to the development of the productive forces – which are in tension with environmentalist beliefs and values. A central task of this book, then, will be to examine such arguments and to suggest that in fact Marx's thinking in these and other areas can be interpreted in ways which are compatible with a recognition of environmental constraints and which offer promising insights into the dynamics of the interaction between humans and nature.

A further qualification of my original formulation should be noted at this point: the confrontation between Marxism and ecologism is not entirely a hostile one. There are, as we shall see, important aspects of Marx's theory – in particular his view of the way in which human societies are dependent upon and moulded by natural conditions, and his concern for a wider range of values than those expressed in the market values of commodities – which mirror the concerns of many green theorists, and this provides another reason why we may reasonably hope that an investigation of Marxism's ecological implications will be a fruitful exercise.

Of course, mine is not the only treatment of these issues. Many of the others are discussed in the following chapters, where the points of agreement and differences of interpretation will emerge. However, there is a variety of levels at which the confrontation (if that is the word) between Marxism and ecologism can be studied, and it is worth saying something further about the place occupied by this book. This study is located at the more theoretical end of the spectrum, addressing philosophical questions of value and forms of explanation, and the most general questions of human nature and of humans' relation to nature, which I take in some sense to be foundational for more concrete and applied forms of investigation, for example detailed investigations of particular economic and political arrangements, political movements and so on, such as those published in the journal *Capitalism, Nature, Socialism*. The focus on normative and explanatory issues in the interpretation of Marx raises the question of how the present account stands in relation to the Analytical Marxism of theorists such as G. A. Cohen, Jon Elster, John Roemer and Erik Olin Wright. If by 'Analytical Marxism' is meant a style of investigation which examines and seeks to clarify problematic concepts and claims in Marx and to interpret or reconstruct his theory in a way which is philosophically defensible, then this is something to which any theoretical defence of Marx must aspire. At a more substantive level the present work expresses some reservations about the methodological individualism (at least in the strong form supported by Elster) that is often said to characterise Analytical

Marxism. As we shall see, however, differences exist within Analytical Marxism itself (notably between Elster and Cohen) over the nature and significance of this doctrine. More generally, the present work owes much to the reconstructive efforts of Analytical Marxism, and in particular to Cohen, whose functional interpretation of historical materialism forms the starting-point for my own account.

The book is structured as follows. The first chapter lays the groundwork, or more precisely it determines the scope and approach of what is to follow, by considering how ecological or environmental problems should be conceived. I discuss the ways in which such problems may be distinguished from others faced by society, and I consider the normative criteria according to which they are judged to be problems, rejecting the 'ecocentric' perspective associated with Deep Ecology and arguing for a form of anthropocentrism, albeit a broader and more nuanced form than is often encountered in the literature.

In chapter 2 I examine one of the key concepts of green politics and environmental literature generally: the concept of natural limits, and, in particular, limits to population and economic growth. This concept is Malthusian in its origins, and it is sometimes argued that Marx and Engels's critique of Malthus constitutes a refusal to accept the existence of environmental limits. I argue against this view, however, and draw upon their critique to suggest the need for a more rounded approach; an approach which recognises that environmental limits are not purely natural, and acknowledges the role that social and technological factors play in their formation.

The task of the remaining chapters is to consider whether Marx's theory of historical materialism is consistent with the recognition of environmental limits, thus understood. Chapter 3 prepares for this by examining the methodological precepts which guide Marx in the construction of his theories, precepts which have been criticised as inadequate for the investigation of ecological phenomena but which in fact anticipate much that is contained in the environmentalists' own methodological speculations. Chapter 4 argues that a recognition of human dependence upon nature is central to Marx's historical materialism; thus he has every reason to accept the reality of environmental limits and to allow for them in his theory of social development. It is often said, however, that this is contradicted by Marx's commitment to the development of the productive forces. Chapter 5 – arguably the core chapter of the book – challenges this contention, offering an interpretation of the development of the productive forces, consistent with the role that it plays within Marx's theory, which – far from implying the transgression of environmental limits – allows that the avoidance or amelioration of ecological problems may serve as a criterion for that development. Since the factors that actually shape technological development may differ according to the intentional structures produced by prevailing relations of production it follows that this ideal of an ecologically benign development of the productive forces may serve as the basis for an ecological critique of existing society and a motivation for change. One of the reasons for Marx's commitment to the development of the productive forces is that such development is necessary in order to achieve the satisfaction of human needs that Marx sees as a condition for the establishment of a communist society, and in chapter 6 I continue and (I hope) conclude the argument by examining Marx's account of human needs, and its ecological implications.

1 Ecological problems: definition and evaluation

In order that we may investigate the ability of Marxism to deal with eco-logical problems – the extent to which Marxist explanations and predic-tions are affected by the existence of such problems and the potential of the theory to explain and offer responses to them – we need to have some idea of what these ecological problems are. Without that we will be unable to identify what is required of the theory or to assess the accounts of ecolog-ical problems given by Marx and Engels. In this first chapter I will there-fore consider the following two questions, which are central to the enterprise of defining ecological problems.

(i) What distinguishes that subset of problems faced by society that are referred to as ecological problems?
(ii) What are the values or moral perspectives that lead to these phenom-ena being regarded as problems?

There is a difficulty involved in attempting to define a phenomenon prior to putting it in a theoretical context, since part of the function of a theory is to provide us with a set of terms with which to characterise the phenomena which the theory addresses. As Hegel put it: 'A preliminary attempt to make matters plain would only be unphilosophical, and consist of a tissue of assumptions, assertions, and inferential pros and cons, i.e. of a dogmatism without cogency, as against which there would be an equal right of counter-dogmatism.'[1] The point is that it is only in the context of a theory which attempts to understand an issue that we can decide whether a particular way of structuring or defining that issue is a good one. Without such a theory, Hegel maintains, we can have no good reason for preferring one definition to another and are therefore vulnerable to the charge of dog-matism. It is evident, however, that some sort of preliminary definition is

[1] Hegel 1975, p. 14.

required, in order to determine the scope of enquiry, and that to proceed without it would also be to open oneself to the charge of dogmatism, since a definition of ecological problems generated from within a particular theory (e.g. Marxism) will inevitably exclude from consideration any problems to which that theory's conceptual scheme renders it blind. I will consider some specific claims about Marxism's supposed blindness to certain aspects of ecological problems in the subsequent chapters. For now, however, the task is to give a preliminary account of what those problems are. In order to avoid the charge of dogmatism, and in particular the charge that my responses to the above questions exclude aspects or examples of ecological problems that are awkward for Marxism, I will draw upon a range of environmental literature and attempt to address the questions by considering intuitions that are widely shared and arguments that are accessible to all participants in the debate and not just adherents of a particular perspective. Thus, while I will at times relate this account to Marxism, I will not be presenting a specifically Marxist account of ecological problems.

1.1 What are ecological problems?

It is sometimes held that the term 'ecology' is properly used to refer to a branch of biology – that which deals with the relations between organisms and their environments – and that it is somehow debased when it is used in connection with environmental campaigns, green parties, and so on. This thought leads some writers to avoid the term 'ecological problem' in relation to the objects of such campaigns, and to write instead of 'environmental problems'. Others – John Passmore, for example – do refer to 'ecological problems', but qualify this as a loose or extended usage of the term.[2] Others again use the term 'ecology' to signify an outlook that is 'deeper' or more radical or fundamentalist in its view of the relation between humans and their environment than mere 'environmentalism'.[3]

It is true that the application of the term 'ecology' to humans takes it beyond the exclusive realm of biology, since (as we shall see) the relation between humans and their environment is importantly mediated by social and technological factors whose study is beyond the scope of that science, and it is true also that the terms 'ecological' and 'environmental' carry dif-

---

[2] Passmore 1974, p. 43.
[3] This is apparent, for example, in the name of the so-called Deep Ecology movement, and also in Andrew Dobson's (1990, p. 13) distinction between 'ecologism' and 'environmentalism'.

ferent associations, the former tending to place more emphasis than the latter on the holistic or systemic aspect of the organism–environment relation. However, these facts do not force us to conclude either that the human–environment relation falls outside the proper realm of ecology, or that there is any difference in the core meanings of the terms 'ecological' and 'environmental' as applied to human problems. I will therefore use the terms 'ecological problem' and 'environmental problem' interchangeably in recognition of the fact that, since humans are organisms, their relation to their environment falls properly within the subject-matter of ecology as stated above. This usage is increasingly reflected in the practice of academic ecology which, according to one of its practitioners, 'has grown from a division of biological science to a major interdisciplinary science that links together the biological, physical, and social sciences'.[4] It follows that any debasement that the term 'ecology' does undergo in connection with its use in relation to 'ecological problems' arises not from its extension to humans and beyond pure biology, but from the particular content that is ascribed to the human–environment relation in its name.

The fact that ecological or environmental problems are not wholly a matter for natural science highlights a difficulty apparent in attempts to define these problems as distinct from others faced by society. As might be expected from the account of the subject-matter of ecology given above, such definitions typically depend upon a distinction between man or society on the one hand, and the environment or nature on the other. Passmore, for example, states that 'a problem is "ecological" if it arises as a practical consequence of man's dealings with nature'.[5] This distinction, however, lacks a clear and unambiguous sense. Reliance on an unexamined notion of nature is likely to prove particularly problematic in considering how Marx and Engels did or could respond to ecological problems, given their insistence that humanity is a part of nature and that nature is transformed or 'humanised' by human activity.[6] More generally, the vagueness of 'nature' is problematic in defining ecological problems, since these problems occur typically (though not necessarily) in situations where the environment *has* been transformed by human activity.

This vagueness in the notion of an ecological problem has sometimes been exploited in order to play down the ecological challenge to Marxism

---

[4] Odum 1975, p. 4.

[5] Passmore 1974, p. 43. Passmore's definition is also adopted by Robin Attfield (1991, p. 1) and, provisionally, by Reiner Grundmann (1991b, p. 23).

[6] E.g. Part of nature: *Economic and Philosophical Manuscripts*, pp. 67, 136; *The German Ideology*, pp. 42, 48. Transformation of nature: *The German Ideology*, p. 62; *Capital*, vol. I, pp. 283–4; *Dialectics of Nature*, p. 172.

by denying the novelty of ecological problems and asserting a continuity between these and the sorts of problems that were addressed by classical Marxism. For example, Hans-Magnus Enzensberger argues that the problems to which twentieth-century environmental movements address themselves are essentially no different from the effects of nineteenth-century industrialisation, which 'made whole towns and areas of the countryside uninhabitable' as well as endangering life in the factories and pits:

There was an infernal noise; the air people breathed was polluted with explosive and poisonous gases as well as with carcinogenous [*sic*] matter and particles which were highly contaminated with bacteria. The smell was unimaginable. In the labour process contagious poisons of all kinds were used. The diet was bad. Food was adulterated. Safety measures were non-existent or were ignored. The overcrowding in the working-class quarters was notorious. The situation over drinking water and drainage was terrifying. There was in general no organized method for disposing of refuse.[7]

What is different now, Enzensberger suggests, and what has led to the emergence of the environmental movement, is not the intrinsic nature of the problems but their universalisation: the fact that they now impinge upon middle-class interests. Enzensberger's view is thus at odds with the view of many greens that environmental problems *are* qualitatively different from (other) social problems in such a way as to create the need for a new political ideology with distinctive proposals for restructuring the whole of political, social and economic life.[8] Gus Hall, also writing from a Marxist perspective, acknowledges that the environmental crisis is 'not just another problem, but a qualitatively different one', requiring 'a radically new approach'; but nevertheless, like Enzensberger, he compares environmental problems with what he labels 'the oldest and most brutal of capitalism's crimes', the deaths resulting from workplace conditions which have 'been going on in the factories and mines for over a hundred years'.[9]

Many of the problems described by Enzensberger can plausibly be classed as ecological or environmental problems. Other writers, however, have drawn the boundary even more widely. Joe Weston, for example, includes street violence, alienating labour, poor and overcrowded housing, inner city decay and pollution, unemployment, loss of community and access to services, and dangerous roads as environmental issues.[10] The fourth item on this list, and perhaps the third, may reasonably be counted as environmental problems, but while the other items may be causes or

---

[7] Enzensberger 1974, pp. 9–10.   [8] Dobson 1990, p. 3.   [9] Hall 1972a, pp. 68, 34.
[10] Cited in Pepper 1993, p. 437.

effects of environmental problems, to count all of them as being themselves environmental problems, as Weston does, is to discard normal usage in a way which deprives the concept of its specificity.

Given that a boundary narrower than Weston's is needed, the problem remains of how it is to be drawn. An individual exists within a whole series of overlapping and nested environments – home, workplace, street, town, country, etc. – each of which has both physical and social components. In a sense, therefore, problems arising in relation to any of these environments could (following Weston) be classed as environmental problems. However, we are concerned with the sense in which 'environmental problem' is equivalent to 'ecological problem', and it is clear (from the discussion of this equivalence above) that ecology is concerned with the relation of the organism to its *physical* environment. Further, as Odum notes, ecology is primarily concerned with levels of organisation beyond that of individual organisms, i.e. with populations and (biotic) communities.[11] Perhaps, then, rather than looking at the individual's relation to his or her environment, which in its broadest sense will include the social environment made up of other human beings and their activities, we should define ecological problems as those concerning the relation between society as a whole and *its* environment – the non-human world, or 'nature'. This brings us back to Passmore's suggestion that ecological problems be defined as those which arise from human dealings with nature. Whatever its faults, this definition does capture the intuition that street crime and the disintegration of communities, for example, are not in themselves ecological problems, and that the workplace conditions referred to by Enzensberger and others fall into a grey area at the boundary of the concept. The workplace is an area in which humans encounter and use materials drawn from non-human nature, yet not all of the problems arising from that encounter fit easily into the concept of an ecological problem: pollution of the atmosphere and waterways, for example, intuitively fits the concept better than the dangers posed by unguarded machinery. This difference, however, appears congruent with Passmore's definition, in that the problems of pollution are essentially concerned with aspects of the natural environment (the air or water or whatever it is that is polluted) in a way in which the dangers of unguarded machinery are not.

The problem with Passmore's definition, as stated above, is the vagueness or ambiguity of the term 'nature'. If by this we mean 'untouched

---

[11] Odum 1975, p. 4. In ecological terms, 'population' designates a group of individuals of a single kind of organism, while 'community' (or 'biotic community') designates all of the populations of a given area (*ibid.*).

nature', excluding objects that have been transformed by human activity, then we will exclude many if not all of the problems generally regarded as ecological. For, as Engels pointed out, 'there is damned little left of "nature" as it was in Germany at the time when the Germanic peoples immigrated into it. The earth's surface, climate, vegetation, fauna and the human beings themselves have continually changed, and all this owing to human activity . . .'.[12] The disappearance of 'untouched nature' has also been the subject of more recent discussion, most prominently by McKibben in *The Death of Nature*. Many conservationists acknowledge, however, that the environments they seek to conserve are in varying degrees products of human intervention, and this may be rendered consistent with Passmore's definition if we allow that nature may include elements that have been altered by humans. Here, though, there is a danger of including too much, since everything is 'natural' at least in being comprised of materials that originate in nature and are subject to its laws. Thus if we stretch the concept of nature too much we will be unable to exclude any of the problems facing society from the realm of the ecological. One writer unwittingly illustrates the absurdity of such an account by arguing that, since humans are a part of nature, 'man's works (yes, including H-bombs and gas chambers) are as natural as those of bower birds and beavers'.[13] I say that this account of nature is absurd because, like Weston's list of environmental problems, it is so broad as to deprive the concept under consideration of any specificity. What it indicates, however, is that short of 'untouched nature' there is no clear boundary between what is natural and what is not. Naturalness appears to be a matter of degree, and the concept of ecological problems, if it is defined in terms of nature, will be correspondingly vague.

As a characterisation, in broad terms, of what is generally understood by the phrase 'environmental problem', Passmore's definition is useful. No-one would dispute that environmental problems are to be understood as involving the relation between humans and nature. What must be emphasised however, and is illustrated by the preceding paragraphs, is that such a definition does not provide for a rigorous distinction between environmental and other problems faced by society. The particular characteristics of environmental problems and the implications of such problems for political theory cannot be derived from a formal definition of environmental problems or an abstract distinction between the concepts of 'humanity' and 'nature', but must be based upon a theoretical account of the actual relation between human beings and their natural and man-made environment.

---

[12] *Dialectics of Nature*, p. 172.   [13] Watson 1983, p. 252

In order to provide the framework for such an account, and to provide a further indication of the scope of this study, I will in the next section approach the problem of characterising ecological problems from a different angle, by examining the categories of phenomena which various writers have put forward as constituting the broader category of environmental or ecological problems.

1.2 Categories of environmental problem

Despite the lack of a rigorous analytical definition of what constitutes an environmental problem, there is a fair measure of agreement about the actual types of problem which fall within this category. Reiner Grundmann compiles a list of environmental 'phenomena' drawn from the 1987 report of the World Commission on Environment and Development (The Brundtland Report) and from Passmore.[14] From the former:

(1) pollution (air, water);
(2) depletion of groundwater;
(3) proliferation of toxic chemicals;
(4) proliferation of hazardous waste;
(5) erosion;
(6) desertification;
(7) acidification;
(8) new chemicals

and from the latter:

(9) pollution;
(10) depletion of natural resources;
(11) extinction of species;
(12) destruction of wilderness;
(13) population growth.

Grundmann argues that this list of phenomena can be reduced to three categories: pollution, depletion of (renewable and non-renewable) resources, and population growth. The last of these is the most controversial, so let us consider it first.

Population growth, Grundmann argues, can be an ecological problem in two senses:

[14] Grundmann 1991b, p. 13; World Commission on Environment and Development 1987, p. 10; Passmore 1974, p. 43.

First, it can be seen as leading to ecological problems such as pollution or depletion of resources, because an increasing population might require more intense exploitation of resources or more technological development with pollution as a side-effect. Second, it can be seen as an ecological problem *per se*, i.e. the increasing number in a specific place may be detrimental to human well-being. Taken in the first sense it is a cause of, taken in the second sense it is an instance of, an ecological problem.[15]

Neither of these statements, however, shows what Grundmann intends. The claim that population growth is a *cause* of ecological problems does not entail that population growth *is* an ecological problem. Interestingly Grundmann does not include other alleged causes of environmental problems such as economic growth or technological development in his list. Secondly, if increasing population were detrimental to human well-being this would show it to be a social problem but would not in itself show it to be an environmental or ecological problem. Even *The Limits to Growth* and *A Blueprint for Survival*, two publications from the early 1970s commonly described as neo-Malthusian because of their assumptions of exponential growth and severe warnings of the dangers of population growth, treat population growth as a cause rather than an example of environmental problems.[16]

What about the other steps that Grundmann makes in reducing to three categories his classification of environmental problems? Grundmann rightly includes 1, 3, 4, 7 and 8 within the category of pollution. He includes 11 and 12 (and presumably 2, though this is not stated) within the category of resource depletion. Problems of food supply (whether arising from population growth as suggested by the *Limits to Growth*, or from other causes) may also be included within this category. More contentiously, Grundmann discards erosion (5) and desertification (6) from the list of environmental problems on the grounds that these are natural processes and are 'interesting in our context only insofar as they are caused by human intervention'.[17] Insofar as this is true, he argues, they can be subsumed under the depletion of natural resources. It is unclear, however, why Grundmann thinks depletion of resources should be regarded as an environmental problem only when it is caused by human intervention.[18] If resources are defined as materials instrumental to human ends,[19] it follows that in order to count as an example of resource depletion the phenomenon in question must, at least potentially, have an *effect* upon human activ-

---

[15] Grundmann 1991b, p. 14.   [16] Meadows *et al.* 1974, p. 23; Goldsmith *et al.* 1972, p. 3.
[17] Grundmann 1991b, p. 14.
[18] It is equally unclear why Grundmann thinks that 2 and 11 cannot happen naturally.
[19] This definition will be qualified in the conclusion to this chapter.

ity, but this tells us nothing about its *cause*. It follows from the fact that we are considering resource depletion as an ecological problem that its causes must consist at least partly of natural or non-human factors, and we have seen no reason to suppose that they cannot be wholly natural.

If we reject population growth as a category of environmental problem, we are left with two categories: depletion of (renewable and non-renewable) resources and pollution. There is one further category which should be considered. This is given by Robin Attfield as 'the endangering of the life-support systems of the planet', and in the *Blueprint for Survival* as the 'disruption of ecosystems'.[20] The introduction of this category reflects what is often referred to as the systemic or holistic nature of ecological problems: the interconnection of different environmental factors and problems.

It might plausibly be argued that this last category is unnecessary, since the disruption of natural systems qualifies as an ecological problem only to the extent that it involves pollution and/or the depletion of resources. On the one hand, disruption of natural systems may be a *cause* of pollution or resource depletion, in which case it would, like population growth, be simply a cause of ecological problems and not an ecological problem in its own right. On the other hand, disruption of natural systems might be counted a case of resource depletion in its own right if 'resource' is defined broadly to include anything which serves the interests or purposes of humans or other creatures, making the disruption of natural systems merely a subcategory of resource depletion. I suggest, however, that there is reason to resist such a reduction. For one thing, such a reduction strains ordinary usage: global warming due to the greenhouse effect is a key example of the disruption of natural systems, and is usually thought of as an example rather than merely a cause of ecological problems, but it can only be subsumed under resource depletion with some artificiality since we would not normally speak of the global climate system as a resource. More importantly, the proposed reduction loses a real and significant distinction between two different kinds of problem: the decline in reserves of a quantitatively measurable substance, such as oil, water or fish stocks, and the breakdown of an interconnected system, such as a marine ecosystem or the global climate system. It is preferable, therefore, to retain disruption of natural systems as a separate category.

---

[20] Attfield 1991, p. 1. Attfield's complete list of ecological problems is: pollution, diminishing natural resources, the increasing size of the human population, the destruction of wildlife and wilderness, loss of cultivable land through erosion and the growth of deserts, and the endangering of the life-support systems of the planet. The *Blueprint* quotation is from Goldsmith *et al.* 1972, p. 3.

1.3 Values and the environment

In the previous two sections I have considered the scope of the concept of
an ecological problem and how we might distinguish these problems from
others faced by society. In this section I will consider the values or ethical
perspectives in the light of which these phenomena are viewed as problems.

Environmental ethics is a branch of applied philosophy which has
developed in recent years, alongside the growth of the environmental
movement, in order to address this issue. As the editor of one collection of
essays puts it: 'What essentially interests us as philosophers is the question: *why* ought we to be concerned with the environment? What moral
principles underlie such a commitment?'[21] Much of the debate in environmental ethics concerns the opposition of two broadly conceived evaluative
frameworks: on the one hand, 'anthropocentrism', which holds that only
humans are worthy of moral consideration for their own sake and that we
should preserve the environment solely for the sake of the humans who
inhabit it, and, on the other hand, approaches described variously as 'biocentric', 'ecocentric', even 'cosmocentric', which ascribe moral considerability to some or all of non-human nature.[22] As the plurality of terms
suggests, a variety of nature-centred ethics can be found, differing in the
range of non-human entities held to be morally considerable. The most
fiercely debated distinction, however, is that between anthropocentric perspectives on the one hand, and non-anthropocentric perspectives ascribing moral considerability to at least some non-human entities on the
other.[23] A widespread view among environmental ethicists is that it is the
anthropocentrism dominant in contemporary societies that is responsible

[21] Dower 1989, p. vi.   [22] The term 'cosmocentrism' is from Bures 1991.
[23] It has been suggested that 'anthropocentrism' is an inappropriate name for the particular
evaluative perspective in question here, since *all* ethical doctrines are unavoidably
anthropocentric in the sense that they are human perspectives and that their judgements
about what is valuable for its own sake must, however widely the boundary is ultimately
drawn, begin from our understanding and experience of what is valuable to us. The substantive doctrine that only humans possess intrinsic value is better captured, it is suggested, by terms such as 'speciesism', 'human chauvinism' or 'human racism'. See
Hayward 1997, Eckersley 1992, pp. 55–6 and Eckersley 1998. My response to this is (i)
simply to stipulate that I am here following the common practice of using the terms
'anthropocentrism' and 'non-anthropocentrism' to refer to the substantive ethical doctrines set out above, and not to make any claims about the perspectival nature of such doctrines, and (ii) to note that the alternatives mentioned above are unsatisfactory in that they
suggest *unwarranted* discrimination in favour of our own species, and hence appear to beg
the question of whether an 'anthropocentric' perspective might be justified. (If a substitute
term for 'anthropocentrism' is insisted upon, then 'human exclusiveness' might be a better
candidate.)

for their ecological crises, and that a shift to a non-anthropocentric ethic is necessary for their solution.[24]

Marx himself would have had little patience either with this view or with the argument more generally, given his well-known aversion to morality and moralising. Marx's views in this area have been the subject of extensive debate, which cannot be adequately addressed here.[25] I will suggest, however, that his hostility to moral discourse need not be as inimical to an environmental ethic as it may at first seem.

Marx's aversion to moralising relates primarily to what he saw as the efforts of Utopian Socialism to convince the ruling class by moral argument of the injustice of capitalism and the moral superiority of socialism. Thus, in the *Communist Manifesto* Marx and Engels write that socialists of this kind 'consider themselves far superior to all class antagonisms' and that they 'habitually appeal to society at large, without distinction of class; nay, by preference to the ruling class. For how can people, when once they understand their system, fail to see in it the best possible plan of the best possible state of society?'[26] Marx and Engels's scorn for such moralising stems from their belief that moral consciousness, as a part of society's superstructure, is conditioned by its economic foundation in such a way that it will tend to reflect the interest of the dominant class in maintaining the status quo. Seen in this light Marx's hostility to morality may be interpreted not as a rejection of moral criticism *per se*, but as an assertion of its limited usefulness as a tool of social change. So, for example, Marx's observation in *Capital* that, since exploitation arises out of the purchase of labour power at a price equivalent to its value, it is 'a piece of good luck for the buyer but by no means an injustice towards the seller'[27] can be interpreted not as a straightforward acceptance of the justice of capitalism, nor a rejection of the possibility of a moral critique, but as an assertion that capitalism is acquitted of injustice according to the (bourgeois) conception of

---

[24] Indeed, some define environmental ethics in such a way that only a non-anthropocentric ethic can qualify. For example Elliot and Gare define an environmental ethic as a systematic ethic which 'allows that future generations, nonhuman animals and nonsentient nature are all morally considerable' (1983, p. x). It seems to me, however, that – whatever the merits and demerits of the two perspectives – it is unhelpful to *define* environmental ethics in these terms. 'Environmental Ethics' is better viewed as the name of the discipline which addresses questions about the value of non-human nature and the reasons there are for preserving it, and within which different answers may be proposed. Otherwise there is a danger that the debate about the most satisfactory form of environmental ethic will be decided by means of linguistic stipulation, yielding a hollow victory for non-anthropocentrism and leaving open the question of how best to understand the value of non-human nature.

[25] See, for example, Lukes 1985 and the responses in McLellan and Sayers 1990. A valuable survey of the debate is provided by Geras 1989.

[26] *Manifesto of the Communist Party*, p. 60.   [27] *Capital*, vol. I, p. 301.

justice that it gives rise to and makes dominant. And although Marx thinks that this bourgeois conception of justice (as equal exchange in the sphere of circulation) will tend to predominate in a bourgeois society, and is at times dismissive of the possibility of bringing any other conception of justice to bear on the process of capitalist exploitation, he does implicitly allow and even make use of such alternative conceptions. This is evidenced by his parallel insistence on the *in*equality of the labour relation considered from the point of view of production, whereby the worker receives less than the value he creates and is therefore, in Marx's words, a supplier of 'unpaid labour' and a victim of 'robbery', 'theft', 'embezzlement' and 'extortion'.[28] Moreover, though Marx holds that bourgeois principles of justice cannot be superseded until material conditions allow for a transformation of the society that spawned them, he nevertheless holds that they can, along with the transitional socialist principle, 'to each according to his work', be judged and found wanting by comparison with the 'higher' communist principle, 'to each according to his need'.[29]

I will return to consider the ecological implications of the needs principle in the final chapter. The point to be made here, however, is that Marx need not be interpreted as denying the truth or objectivity of evaluative statements critical of capitalism; his point is that even if they are true they will not gain general acceptance and their promotion is therefore ineffective as a means of bringing about social change. For Marx, therefore, the fundamental task of a theorist engaged in criticism of existing society is not the promotion of one or another set of moral beliefs, but the analysis of the various interests and the social structures which underlie them, in order to identify and promote potential agencies of change.

Interpreted in this way, Marx's scepticism about morality may serve as a warning against regarding ecological problems as simply the result of a wrong set of values, to be rectified by the promulgation of a new ethic without considering the interests and structures underlying those values. It should also caution us against expecting to find a ready-made environmental ethic in his works. It does not, however, imply that evaluative issues should be ignored. Firstly, Marx overstates the ineffectiveness of moral argument. Even if truths about the injustice of capitalism are inaccessible to those who benefit from it (which is itself an overstatement), such truths may nevertheless have an important motivational effect upon the victims of that injustice. Marx's belief seems to be that their interests alone will suffice to motivate them, but this ignores the degree to which a

---

[28] See Geras 1989, pp. 223, 225.
[29] *Critique of the Gotha Programme*, p. 320; cf. Geras 1989, pp. 227–8.

person's motivation may be strengthened by the belief that the objective she is pursuing is not only in her interest but something to which she has a right, and even something she has a duty to pursue. Secondly, we cannot proceed without addressing evaluative issues, since the choice of evaluative perspective will affect what we count as an environmental problem and what would count as its resolution. For example, global warming and the resultant rising of sea levels might, from an anthropocentric perspective, be solved by the development of drought-resistant crops, desalination plants and coastal defences to counteract the threat posed to human interests. Non-anthropocentrists, however, would continue to regard these phenomena as problems so long as they impact upon the other living things with whom we share the environment, threaten species extinction, or undermine the integrity of natural ecosystems. Clearly, therefore, in order to assess whether Marxism can meet the challenge posed by environmental problems we must address the claims of a non-anthropocentric environmental ethic.

Non-anthropocentrism was defined above as the view that moral considerability extends beyond human beings. It is important, however, to qualify this by noting that theories of 'animal liberation', such as those of Peter Singer and Tom Regan, which assert the moral considerability of the 'higher' animals (those that share with humans characteristics such as sentience or the capacity to suffer) are often regarded as being on the anthropocentric side of the divide, or any rate not fully non-anthropocentric. On this view, non-anthropocentric perspectives would, despite the terminology, be more accurately characterised as those for which moral concern extends beyond the interests of individual sentient creatures.[30] Whatever the merits of such a definition, it is true that the extension of moral concern to sentient creatures is, both theoretically and in its practical implications, a less radical and less controversial departure from strict anthropocentrism than its extension to such things as plants, species and ecosystems. In what follows I will therefore assume that an anthropocentric approach can be extended to include sentient creatures other than humans among the

---

[30] A much cited assertion of the incompatibility of animal liberation and a (non-anthropocentric) environmental ethic is Callicott's 'Animal Liberation: A Triangular Affair' (in Callicott 1989). Callicott has since modified his view, as he explains in a preface to the reprint of his article in Elliot 1995. For further discussion of this issue and other references, see Jamieson 1998; Attfield 1991, pp. 179–81; Eckersley 1992, pp. 44–5. The main arguments for not counting animal liberationism as a fully non-anthropocentric ethic are (i) that its method (of arguing outwards by analogy from humans to other species) is anthropocentric, and (ii) that in practice it is committed to policies which are incompatible with the agenda of radical ecocentrism. Strictly speaking, however, neither of these shows that it is anthropocentric in the sense defined above.

objects of moral concern, and will focus on the question of whether moral concern ought to be extended to non-sentient parts of nature.

I will assume further that the burden of proof in this controversy lies with those who wish to extend the sphere of moral considerability beyond sentient creatures. This may be criticised as kind of methodological anthropocentrism, but it seems to me that we have no alternative. The view that ecosystems and their components should be preserved not just for the benefit of the humans or other sentient creatures who enjoy or depend on them, but for their own sake, is highly contentious and therefore in need of justification. My investigation will therefore take the form of an examination and critique of the most common and plausible arguments for such an extension of moral concern. We may begin, however, and set the argument in the context of Marx scholarship, by considering the argument *for* anthropocentrism put forward by Reiner Grundmann as part of a defence of Marx (and the Enlightenment tradition which he sees Marx as representing) against ecological critique from a non-anthropocentric perspective.

1.3.1 Flourishing and moral considerability

Grundmann's argument for an anthropocentric approach is based on the supposition that the non-anthropocentrist must distinguish between states of nature which are 'normal' and thus to be preserved, and states which are 'pathological' and thus to be avoided. Against this, Grundmann objects that 'it is difficult to know what is "normal" for nature', and, more strongly, that this cannot be defined without reference to human interests.[31] Standard accounts of ecological normality in terms of 'balance' or 'diversity' only make sense, according to Grundmann, in relation to human interests. One aspect of Grundmann's argument that might be questioned is his identification of 'normality' as the focus of a non-anthropocentric approach, but this does not affect his overall argument. Ecocentrists must identify some states of nature as being intrinsically better than others, and they typically do so in terms of the flourishing of natural systems; Grundmann's response, more generally stated, is that we cannot make sense of what it is for a natural system to flourish except in terms of the human interests served by that system.

There is, however, an important strand of environmental ethics which denies this assertion. According to this Aristotelian approach we *can* make sense of what it is for non-sentient entities to flourish, and we can there-

---

[31] Grundmann 1991b, p. 24.

fore identify states and conditions that are good for them, independently of human interests. We may say, for example, that a plant 'does well' in certain conditions, or that those conditions are 'good for it', without apparently making any assumption about whether we want it to flourish. We can even make sense of the idea of the good of a non-sentient entity conflicting with human interests: crowded buses, we may say, provide good conditions for the propagation of the flu virus; mild winters are good for greenfly and therefore bad for gardeners.[32] Whatever contributes causally to an object's flourishing is instrumentally good for that object, and whatever constitutes its flourishing is intrinsically good for it. Thus, anything that can be said to flourish can be said to have its own intrinsic goods, or intrinsic values, independent of human evaluation. Such things may thus be designated objects of direct moral concern by a theory which enjoins the promotion or preservation of such goods.[33]

Within this broad framework there is disagreement about the kinds of things that can be said to possess goods of their own. Adherents of a 'biocentric' or life-centred ethic, such as Robin Attfield and Paul W. Taylor, attribute goods of their own only to individual living organisms, whereas 'ecocentrists', such as Baird Callicott, Lawrence Johnson and Holmes Rolston, attribute them to collective or 'systemic' entities such as species, ecosystems, and even the biosphere as a whole.[34] It is arguable that if we do attribute goods of their own to individual non-sentient organisms, then we should attribute them also to such things as species and ecosystems, since we do intuitively seem able to make sense of the idea of such things having their own goods. An ecologist might say, for example, that a species does well when its population is large and stable rather than small and

---

[32] This example comes from O'Neill 1993, p. 22.

[33] The idea that objects having goods of their own are morally considerable is also articulated by some theorists in terms of *interests*, such that objects with goods of their own have an interest in realising such goods and therefore fall under a moral principle enjoining consideration of the interests of others. See, for example, Attfield 1991, pp. 144–5. However, others (e.g. Taylor 1986, pp. 60–71) distinguish between 'having interests' and 'having goods of one's own', limiting the former to cases where a creature has goods which it is conscious of and strives to achieve, while holding the latter to be sufficient for moral considerability.

[34] James Lovelock (not himself a Deep Ecologist but a source of inspiration to many of them) famously characterises the biosphere as a superorganism, named Gaia after the Greek earth-goddess. He is careful to avoid attributing sentience to this superorganism (or at least – perhaps there is a deliberate ambiguity here – he is careful to avoid supposing that we *know* it to be sentient) but nevertheless assumes that we can make sense of the idea of Gaia flourishing or not – and that her flourishing may be incompatible with human flourishing. (Lovelock 1995, p. ix.) Among the writers mentioned above, Johnson (1993, pp. 265–6) takes the biosphere to be a system with interests of its own, distinct from those of its constituent subsystems. Rolston (1994, pp. 25–8) appears to take a similar view with regard to the earth considered as a natural system.

declining, or that an ecosystem is flourishing when it is able to maintain stability despite changes in the wider environment. Attfield and Taylor may appear, therefore, to occupy an unstable middle ground between Grundmann on the one hand, denying that we can identify what it is for *any* non-sentient entity to flourish except by reference to our own interests or preferences, and full-blown ecocentrism on the other.[35] But while Attfield and Taylor contest this view[36] it is not necessary to decide the matter here, because even if we accept that a wide range of non-sentient entities have 'goods of their own' this is not sufficient to establish their moral considerability.

1.3.2 Objections and responses

Non-anthropocentric ethics are often criticised for their propensity to generate moral conclusions that are abhorrent or unworkable – conclusions that require vital human interests to be sacrificed for the good of non-sentient entities.[37] The propensity to generate such conclusions seems particularly pronounced in the case of holistic ethics which view the ecosystem or 'biotic community' as the primary repository of value, and its component parts (human individuals and others) as valuable only insofar as they contribute to the flourishing of the whole. Such views have been labelled 'environmental fascism' by Tom Regan.[38] But it is not only holists who are vulnerable to such an argument, as Attfield acknowledges:

[35] Cf. Eckersley's (1992, p. 47) observation that biocentrism is not a major stream of environmentalism since non-anthropocentric theorists have tended to gravitate towards animal liberation on the one hand, or deeper ecocentric approaches on the other.

[36] They argue that the interests or goods of collective entities such as species, insofar as they exist at all, are reducible to those of their present, or present and future members. See Attfield 1991, pp. 150–1; Attfield 1995, pp. 24–5; Taylor 1986, pp. 69–70; and, for a rejoinder, Johnson 1993, p. 183. Attfield further suggests (1995, p. 26) that the patterns of growth of collectivities such as species could only count as analogues to those of individuals, and hence ground notions of flourishing and interests, if some counterpart could be found for the genetic determination of individuals' capacities – but surely these capacities and patterns *are* determined by the genetic make-up of their constituent members, and in any case it is unclear to me why the mode of determination should be regarded as important.

[37] See, for example, Grundmann 1991b, p. 24. See also Bookchin's critique of Deep Ecological misanthropy, cited in Low and Gleeson 1998, p. 144.

[38] Regan 1988, pp. 361–2. Regan's primary target is Aldo Leopold's famous 'land ethic' (proposed in his *Sand County Almanac*), which has influenced writers such as Callicott and Rolston, and which holds that actions are right when they contribute to 'the integrity, stability, and beauty of the biotic community' and wrong otherwise. Note, however, that while this quotation tends to support the holist view attacked by Regan, others of his formulations suggest an *extension* rather than an abandonment of human-centred ethics, which would not deny the moral considerability of humans and other individual creatures. For discussion of this and of the charge of environmental fascism, see Johnson 1993, pp. 175–8 and Attfield 1991, pp. 157–9.

The objection may . . . be expressed as follows. If plants (or bacteria) have any more-than-negligible moral significance, then in their millions their interests must sometimes outweigh those of individual humans or other sentient beings; but this flies in the face of our reflective moral judgements, and should thus, short of compelling reasons, be rejected.[39]

Attfield's response is that the moral significance of non-sentient entities may be so small, compared with that of sentient creatures, as to make a difference to the choice of action only when considerations relating to sentient interests are very finely, even perfectly, balanced.[40] This response meets the main thrust of the objection, removing the intuitively abhorrent consequences of a non-anthropocentric ethic, but leaves two problems for the non-anthropocentrist.

The first is that a theory which balances the goods of sentient and non-sentient things in this way will differ very little in its practical prescriptions from one which limits moral considerability to humans and other sentient creatures, and will therefore pose less of a challenge to established political theories, including Marxism, than many non-anthropocentrists intend. For this and other reasons it is a solution that many of them will resist.[41] Secondly, in the kinds of case under consideration, what may be considered morally repugnant is not simply the suggestion that the goods of non-sentient things can *outweigh* important human interests, but that they count for anything at all. We may think, for example, that the 'interests' of the AIDS virus are not simply outweighed by human interests, but rather, that the fact that something is good for the virus is no reason at all, not even a defeasible or *prima facie* reason, for promoting that thing.[42] If we accept this thought then we must conclude that something's having a good of its own is not, as the Aristotelian argument supposes, a sufficient condition for moral considerability.

What the intuition just described highlights is that there is a logical gap between the claim that something has goods of its own, and the claim that

---

[39] Attfield 1991, p. 154.

[40] *Ibid.* Attfield draws a contrast between moral *significance*, which is a matter of degree, and moral *standing*, which is not. Moral standing, according to Attfield, is possessed by all entities which have a good of their own, and indicates that the goods of that entity carry some moral weight. The moral significance of an entity is the amount of moral weight which its goods have in comparison with the goods of other kinds of entity.

[41] See, for example, Taylor 1986, pp. 269–70.

[42] This point is analogous to the following objection to utilitarianism. The utilitarian will say that sadistic torture is wrong because the pleasure obtained by the torturer is more than outweighed by the suffering of his victim, but what ought to be said is that the torturer's pleasure is not a reason in favour of committing the torture at all. Both Attfield and the utilitarian (whose analysis of the torture case is defended in Attfield 1995, pp. 34–5) reach what is intuitively the right answer, but arguably for the wrong reasons.

it is morally considerable or has moral standing. The former is a factual claim, that the object in question has a natural potential or a tendency towards the achievement of certain 'ends', relative to which it may be said to flourish or not.[43] The latter, on the other hand, is a normative claim, that moral agents ought or ought not to treat it in certain ways. As Taylor notes: 'One can acknowledge that an animal or plant has a good of its own and yet, consistently with this acknowledgement, deny that moral agents have a duty to promote or protect its good or even to refrain from harming it.'[44] O'Neill similarly observes: 'That Y is a good of X does not entail that Y should be realised unless we have a prior reason for believing that X is the sort of thing whose good ought to be promoted.'[45] In other words the non-anthropocentrist must show not only that it makes sense to speak of non-sentient things having 'goods' or 'interests', but also that these 'interests' are morally significant ones which we ought to promote.

How may such a thing be argued? If it is accepted that we have no way of deducing normative conclusions directly from factual (non-normative) premises then it is clear that any such argument must appeal to shared moral beliefs. One strategy is to seek analogies between things that are agreed to be morally considerable and those whose moral considerability is in dispute. This is the method used by Singer to argue for the moral considerability of sentient animals: they like us have interests – in the avoidance of suffering if nothing else – and consistency demands that we treat those interests no less seriously than the similar interests of humans. Singer, however, resists any further extension of moral considerability, articulating the intuitive and widely held view that since nothing can matter to a creature which is incapable of experiencing anything, it cannot matter morally what we do to it except insofar as it affects the interests of sentient creatures.[46]

But while this restriction on the range of moral considerability seems obvious to many, others disagree. Attfield, for example, sees an analogy (albeit a weak one which would confer only a limited moral significance) in the fact that the capacities whose fulfilment constitutes the flourishing of non-sentient entities – such as growth, respiration, self-preservation and reproduction – are ones which they share with us.[47] Taylor also appears to use an analogical form of argument when he defends his biocentric ethic as the only one consistent with the 'biocentric world view', which emphasises properties and relationships we have in common with other organisms, including our dependence on biological and physical conditions for

---

[43] This, at least, or something like it, is what the claim means when applied to non-sentient things.   [44] Taylor 1986, p. 72.   [45] O'Neill 1993, p. 23.
[46] See, for example, Singer 1993, p. 277.   [47] Attfield 1991, pp. 153–4, 205.

survival, the fact that we have goods of our own and a capacity to realise them, our common evolutionary origin, and our dependence upon the healthy functioning of the biosphere. The problem with any such argu-ment, however, is that for any proposed widening of the moral community – whether to animals, living organisms or self-regulating systems – we will find both analogies and disanalogies between humans and the wider group.[48] The question is: which ones are relevant to the question of moral considerability? And here Singer's answer again seems plausible: the 'interests' or goods that matter morally are those that matter from the point of view of the entity in question. Thus sentience, or possession of a point of view, remains a necessary condition of moral considerability, without which other analogies are irrelevant.

Taylor attempts to overcome this barrier by ascribing a point of view to whatever has a good of its own. He characterises living organisms gener-ally as 'teleological centers of life', each 'striving to preserve itself and realize its good in its own way', and writes of such organisms that one can achieve 'a genuine understanding of its point of view' and can then 'imag-inatively place oneself in the organism's situation and look at the world from its standpoint'.[49] The language of 'striving' and of 'points of view', however, is metaphorical, and adds nothing of substance to the claim that such things have goods of their own. A tree, for example, does not *try* to reach its potential, and though our view of the world can be informed by an understanding of its genetically programmed tendencies and potentials we cannot literally *look* at the world from its standpoint, since, looking at, or more generally perceiving, the world is not within its capacities.[50]

Another kind of argument for the moral considerability of non-sentient things draws upon our intuitive reactions to particular imaginary scenar-ios. These 'last man' or 'last person' arguments present scenarios in which just one person (whom we may take also to be the last sentient being) is left alive following some disaster, and ask us to judge whether, for example, it would be wrong for that person gratuitously to chop down a tree, or wipe out a species, or unleash a nuclear arsenal that would destroy the remaining life on the planet. The expectation is that we judge it to be wrong, and conclude that the tree, or species etc., has a value, or moral con-siderability, that does not depend on it serving the interests of humans or other sentient creatures.[51] Such arguments, however, present a number of difficulties. Firstly, the intuitions elicited may not be as uniformly

---

[48] *Ibid.*, pp. 154–5.   [49] Taylor 1986, pp. 120–1.

[50] Cf. Singer 1993, p. 277: 'there is nothing that corresponds to *what it is like to be* a tree dying because its roots have been flooded' (my emphasis).

[51] Attfield 1995, pp. 21–2. This style of argument is attributed originally to Richard Routley.

supportive of the non-anthropocentrist case as the authors of the arguments assume. Secondly, intuitive responses to such unfamiliar scenarios are, in any case, likely to be highly theory-dependent and therefore to reflect our background moral theory rather than providing neutral data with which to adjudicate between competing theories. Thirdly, it is difficult if not impossible to separate the situation under consideration from one's own contemplation of it. My own intuitions, for what they are worth, suggest that the value read into such situations is derived from the satisfaction that a sentient being might take in the existence or flourishing of the natural entities in question. It is not necessary to make the mistake of imagining oneself present as an observer in order to intuit such value, though this mistake may be hard to avoid, for it is possible to take satisfaction in the idea of something's existing or flourishing (one's great-grandchildren for example) even though one will never experience the reality. Such arguments fail, therefore, to demonstrate that the objects in question are morally considerable for their own sake.

A further argument to bridge the gap between having goods of one's own and being morally considerable is put forward by John O'Neill, continuing the Aristotelian theme:

> Human beings like other entities have goods constitutive of their flourishing, and correspondingly other goods instrumental to their flourishing. The flourishing of many other living things ought to be promoted because they are constitutive of our own flourishing. This approach might seem a depressingly familiar one. It looks as if we have taken a long journey into objective value only to arrive back at a narrowly anthropocentric ethic. This however would be mistaken. It is compatible with an Aristotelian ethic that we value items in the natural world for their own sake, not simply as an external means to our own satisfaction.[52]

Like many others in the field, O'Neill is taking the ascription of *intrinsic value* to non-human nature as the touchstone of a non-anthropocentric environmental ethic. This is problematic since, as O'Neill points out, the term 'intrinsic value' is used in a variety of different senses, including non-instrumental, non-relational, and objective value.[53] O'Neill's view, however, is

---

[52] O'Neill 1993, p. 24.

[53] One response in the face of such differences is to argue for a particular usage as being more semantically accurate, or more conducive to philosophical clarity than the others. Thus Karen Green (1996), following Christine Korsgaard (1983), argues that 'intrinsic value' is properly understood in contrast to 'extrinsic value', as non-relational value (or, in G. E. Moore's terms, value that depends solely on the intrinsic nature of the thing in question, not its relations to anything else), and that this usage preserves important distinctions that are lost when intrinsic value is equated with non-instrumental value. But while I am in sympathy with this argument, it seems to me that the pervasiveness of other usages is such that clarity will best be served not by stipulating a particular sense of 'intrinsic value' but by substituting terms such as 'non-instrumental', 'non-relational' and 'objective', or at least by qualifying the use of 'intrinsic value' in these or other terms.

that it is the ascription of *non-instrumental* value that determines whether an ethic is anthropocentric: 'To hold an environmental ethic is to hold that non-human beings have intrinsic value in the first sense: it is to hold that non-human beings are not simply of value as means to human ends.'[54]

The claim advanced in the quoted passage is that although the reasons we have for protecting non-human entities are grounded in their contribution to our own welfare, this does not imply that their value is purely instrumental. We do value such things as plants and ecosystems instrumentally, because they contribute causally to our own well-being; but we can also value them non-instrumentally, because their flourishing, and our caring about their flourishing, is a *part* of what it is for us to lead flourishing lives.[55] It seems to me, however, that O'Neill is mistaken to characterise this view as non-anthropocentric, since the non-instrumental values that he plausibly ascribes to non-sentient parts of nature remain *derivative* of human interests.[56] Thus, if it is asked *why* we should preserve a particular species or landscape, the answer will be that we will flourish more, or lead fuller human lives, if we do. It may be that human flourishing will best be served by treating non-human nature *as if* it were morally considerable in its own right, but ultimately, the reasons for adopting such a stance remain fully human-centred.

O'Neill's theory is therefore an anthropocentric one. We may, however, characterise it as a 'broad' anthropocentrism, in contrast with the 'narrow' forms which ascribe only instrumental value to non-human nature. This distinction will be significant later, when we consider the idea of human domination of nature. First, however, we should consider the relation of the so-called 'Deep Ecology' movement to the present discussion of evaluative frameworks.

---

[54] O'Neill 1993, pp. 9–10. Note that O'Neill is here using 'environmental ethic' in the narrow sense criticised in note 24 above, according to which an environmental ethic is by definition non-anthropocentric.

[55] This exposition glosses over some aspects of O'Neill's claim that are in need of clarification. The claim could be construed for example as: (i) the flourishing of X is, in itself, constitutive of my flourishing, irrespective of my attitude towards it; (ii) the flourishing of X is constitutive of my flourishing just if I happen to care about X; (iii) the flourishing of X is constitutive of my flourishing if I care about X, and caring about X-like things is constitutive of human flourishing. The second sentence of the above quotation seems to suggest (i), while the reference in the last sentence to human evaluations of natural items seems to suggest (ii) or (iii).

[56] The distinction between non-instrumental and non-derivative value is discussed (in a different context) in Raz 1986, p. 177. Non-instrumental value is value that a thing has apart from the value of its consequences, or the consequences it is likely to have or can be used to produce. However, a thing's value can also be derivative if it is derived from the value of some whole of which it is a part. (This is sometimes called 'contributive value'.) Non-derivative value (or 'ultimate value' in Raz's terms) is value that is derived *neither* from the value of a thing's consequences *nor* from the value of some whole of which it is a part.

1.3.3 Deep Ecology

The term 'Deep Ecology' is often associated with non-anthropocentric environmental ethics and particularly with forms of ecocentrism that ascribe intrinsic value to ecological wholes such as species and ecosystems.[57] One difficulty with this characterisation is that, according to its main expositors, Deep Ecology comprises a conjunction of views, only some of which are ethical or even philosophical. Thus the Deep Ecology 'platform', formulated by Arne Naess and George Sessions as a summary of the movement's principles, includes claims about the excessive size of the human population, the effects of human interference with the non-human world, and the need for various structural changes, alongside claims about flourishing and intrinsic value.[58] Nevertheless, it might plausibly be claimed, on the basis of this and other central texts, that a non-anthropocentric ethic is an essential component of Deep Ecology. For example, the first principle of the 'platform' states: 'The well-being and flourishing of human and non-human Life of Earth have value in themselves (synonyms: intrinsic value, inherent value). These values are independent of the usefulness of the non-human world for human purposes.'[59] And in his 1973 article, widely regarded as the founding document of the Deep Ecology movement, Naess proposed a principle of 'biospherical egalitarianism', stated in the following terms:

The ecological field worker acquires a deep-seated respect, even veneration, for ways and forms of life. He reaches an understanding from within, a kind of understanding that others reserve for fellow men and for a narrow section of ways and forms of life. To the ecological field worker, the *equal right to live and blossom* is an intuitively clear and obvious value axiom.[60]

Although it is quite unclear what one would have to do to respect the equal rights of, for example, a person, a tree and a blade of grass, it seems clear that Naess is attributing moral considerability to these and other forms of life. In this respect his position would appear to be a form of biocentrism akin to that of Taylor. However, as Naess later makes clear, the prefix 'bio-', and the term 'forms of life', are to be understood as referring not only to life in the strict sense, but also to non-living things such as rivers, landscapes and ecosystems, to 'the ecosphere as a whole' and to every natural item,[61] yielding, so it would seem, an ecocentric ethic.

---

[57] See, for example, Brennan 1988, p. 141; Singer 1993, p. 280.
[58] See Devall and Sessions 1985, p. 70; Sylvan and Bennett 1994, pp. 95–6.
[59] *Ibid.*   [60] Naess 1973, pp. 95–6. Also in Naess and Rothenberg 1989, p. 28.
[61] Naess and Sessions, cited in Sylvan and Bennett 1994, pp. 99–100.

This, however, must be seen in the context of a second aspect of Deep Ecology, as formulated in Naess's founding article. This is a holistic metaphysic, referred to as the 'relational, total-field image', and described, somewhat impressionistically, as follows:

Rejection of the man-in-environment image in favour of *the relational, total-field image*. Organisms as knots in the biospherical net or field of intrinsic relations. An intrinsic relation between two things *A* and *B*, so that without the relation, *A* and *B* are no longer the same things. The total-field model dissolves not only the man-in-environment concept, but every compact thing-in-milieu concept – except when talking at a superficial or preliminary level of communication.[62]

In this passage Naess uses 'intrinsic relations' to refer to what are usually called 'internal relations', that is, relations constitutive of the identity of the relata. Thus, Naess's claim is that the identity of each human being is partly constituted by the relation in which they stand to their environment. Although such a view needs more careful elaboration than Naess provides,[63] it is clear that the thrust of his view is to deny any basic distinction between the self and its environment.[64] Thus, ontological rather than ethical questions become central to Deep Ecology, and the distinction between anthropocentric and non-anthropocentric ethics becomes blurred. If I am one with my environment then valuing my environment is a part of valuing myself, and there is no clear distinction between valuing it for its own sake and for its contribution to my well-being.[65]

Taken literally, however, the denial of a distinction between self and environment is implausible. It is surely possible to conceive of one's self persisting (in however miserable a state) following the destruction of whatever features of the environment one most cares about. This, perhaps, is why Naess himself and other Deep Ecologists have gone on to interpret

---

[62] Naess 1973, p. 95.

[63] In particular it is necessary to specify the aspects of the environment, and the kind of relation between me and them, that are supposedly necessary for the maintenance of my identity. Presumably it is not just 'the environment', but some more specific configuration of it, that is supposed to be constitutive of my identity, since the continued existence of the former is hardly in doubt. And the relationship asserted to be constitutive of my identity could be a purely physical one ('being in' an environment so configured) or an intentional one ('caring for' or 'being aware of' an environment so configured).

[64] Such a view is also implied by Freya Mathews's (1995, p. 143) assertion that '[the universe's] selfhood conditions mine, my selfhood conditions its', and by Warwick Fox's characterisation of the central intuition of Deep Ecology as 'the idea that there is no firm ontological divide in the field of existence' (quoted in Sylvan and Bennett 1994, p. 103). Note however that this differs from Fox's later account, described below.

[65] This, indeed, may be what Naess has in mind when he follows his assertion of biospherical egalitarianism with the claim that the restriction to humans of the equal right to live and blossom 'is an anthropocentrism with detrimental effects upon the life quality of humans' (1973, p. 96).

the holistic aspect of Deep Ecology in other ways, focusing on the conditions of self-realisation rather than those of personal identity. Warwick Fox, for example, sees the cultivation of an expanded, or 'transpersonal', sense of self as the central issue of Deep Ecology (which he therefore renames 'Transpersonal Ecology'). However, he understands this in psychological rather than metaphysical terms, such that it is not the *identity* of the self that is at issue, but its capacity *to identify with* other natural beings.[66] Since 'identification', in this context, would appear to mean something like caring for others in such a way that one seeks their own flourishing, or 'self-realisation', as a part of one's own, this places Fox's version of Deep Ecology in the same 'broad anthropocentric' category as O'Neill's theory, asserting that humans have non-instrumental reasons to preserve natural entities, as constitutive not of their identity but of their well-being.[67]

It would appear, therefore, that Deep Ecology is not characterised by a single distinct evaluative perspective, but rather that it is united in the ethical domain only by a rejection of narrow instrumentalist forms of anthropocentrism, and that this rejection takes various forms, crossing, and sometimes attempting to collapse, the anthropocentric/non-anthropocentric divide. Since I have already considered and found wanting the arguments for a non-anthropocentric ethic, I will now consider further the consequences of a broad, non-instrumentalist anthropocentrism, in relation to Grundmann's defence of Marx.

1.3.4 Broad anthropocentrism and the domination of nature

Grundmann sets out to defend not only the anthropocentrism which he rightly associates with Marx, but also, and more contentiously, the idea of human domination of nature. While many, even of those who defend an anthropocentric ethic, would want to detach it from the idea of domina-

---

[66] For another account of Deep Ecology which emphasises the identification of individuals with their wider environment (and with nature as a whole), see Mathews 1991; Mathews 1995. Mathews, however, does appear to regard the metaphysical claim that my environment is constitutive of my identity and vice versa as necessary to her argument. (1995, p. 150).

[67] It would of course be possible to combine a human self-realisation perspective like Fox's with a non-anthropocentric ethic, if one asserted that we *ought* to cultivate a wider sense of self (wider identification) in order to facilitate the preservation of morally considerable beings. Fox, however, eschews such a justification, offering his self-realisation perspective as an 'experiential invitation' and presenting it as an alternative to the moral perspectives offered by environmental ethics. (See Fox 1990, p. 247; Eckersley 1992, pp. 61–3.) Nevertheless, his perspective does contain an evaluative perspective, which in the absence of a deeper underlying ethic appears for the reasons given to qualify as a broad anthropocentrism.

tion, Grundmann suggests that the two go together, as part of the Enlightenment tradition which he sees Marx as representing.

Far from being an outdated attitude, discredited by the environmental problems over whose emergence it has presided, Grundmann argues, domination of nature is 'a reasonable approach with which we can make sense of the problem and stipulate solutions'.[68] This is because 'domination' is properly understood in terms of human interests and needs, and consequently a society whose transformation of nature brings about ecological problems 'can hardly be said to dominate nature at all'. Grundmann illustrates this by reference to the fable of King Midas who, 'by turning everything he touched into gold, can hardly be said to have "dominated" his citizens, or even his own private life. His power was self-defeating since he was no longer able to feed himself.'[69] Thus, Grundmann sees ecological problems not as a result of the domination of nature but as evidence of its absence.

There are, it seems to me, two problems with this argument. The first concerns Grundmann's understanding of 'domination'. As Grundmann uses the term, to dominate nature is to use it, without moral constraint, in a manner which serves human interests. But in ordinary usage, 'domination' does not always imply an actual furthering of the agent's interests. A manager who dominates his staff, for example, may fail, for just that reason, to obtain the best work from them. In this context, 'domination' appears to refer to the imposition of one's will on someone without regard for that person's interests or autonomy. The fact that this definition cannot readily be carried over to the case of human domination of nature serves to illustrate the problematically metaphorical nature of this usage. Grundmann can reply to this objection by means of a stipulation – that 'domination', in the sense which he wishes to defend and which he associates with the Enlightenment tradition, is to be understood in a way which does imply the successful pursuit of human interests. This he implicitly does, by using 'domination' synonymously with 'mastery', a term which does tend to imply success in one's dealings with that which one masters.

There is, however, another problem with Grundmann's argument, which remains after its reformulation as a defence of human *mastery* of nature. The problem is that the argument depends upon a prior assumption of anthropocentrism. Domination, or mastery, of nature, according to Grundmann, is an appropriate objective because it implies the successful

---

[68] Grundmann 1991b, p. 15.   [69] *Ibid.*, p. 60.

use of nature to promote human interests, and this in turn implies the resolution of ecological problems. However, this presupposes that ecological problems are to be understood in terms of human interests, and it is just this presupposition that is challenged by many opponents of the domination model. The most that Grundmann's argument shows, then, is that *if* the value of nature lies only in its usefulness to humans *then* the domination of nature, properly understood, is an appropriate objective. Now, this presupposition may seem unproblematic since it has been argued above that although Grundmann's defence of anthropocentrism fails to take into account the arguments of its main opponents, these arguments ultimately fail to provide adequate support for a genuinely non-anthropocentric ethic. I will argue, however, that the kind of 'broad' anthropocentrism that emerged from these arguments is one which does not sit happily with the metaphor of domination.

A 'broad' anthropocentrism, as I have defined it, is one which grounds the value of non-sentient nature in its contribution to the value of human lives, but which unlike its narrow counterpart does not view that contribution solely in instrumental terms. A similar distinction is invoked by Grundmann in order to explain the motivation behind non-anthropocentric theories. One of the reasons ecologists have rejected an anthropocentric standpoint, he argues, is that they have equated it wrongly with a perspective of narrow economic short-term self-interest, whereas in fact the human interests underlying evaluations of nature may have a more general, longer-term and aesthetic character.[70] I believe that Grundmann is right in suggesting that a broad anthropocentrism can give an adequate account of the values we find in nature and hence undermine the motivation of non-anthropocentric theories; Grundmann, however, deprives himself of the means of characterising a 'broad' anthropocentrism, and reduces the plausibility of his own theory, by insisting that all valuations of nature, including aesthetic valuations, are instrumental. In this respect O'Neill's theory, despite its misleading non-anthropocentric label, provides a better example of the kind of broad anthropocentrism that I am arguing for. I would add, however, that the range of things recognised by such a theory as contributing non-instrumentally to human well-being need not be limited to the flourishing of things with goods of their own.

We can now see why the metaphor of domination is deficient even from an anthropocentric perspective. Grundmann's argument is that the domination (or mastery) of nature should be understood in such a way that only

---

[70] *Ibid.*, p. 25.

those uses of nature which succeed in furthering human interests count as domination. The problem is that even if we grant him this interpretation, the metaphor of domination emphasises certain ways of relating to nature which, from a broad anthropocentric perspective, do not exhaust the ways in which nature may be valuable to us. One aspect of the broad anthropocentric perspective that Grundmann himself recognises is the aesthetic valuation of nature. This typically takes the form of a passive appreciation and therefore fits uncomfortably with the notion of domination, which suggests an altogether more active, controlling role for human beings. Furthermore, aesthetic valuation is often cited as an example of the way in which an object can be derivatively but non-instrumentally valuable: the aesthetic object, whether art or nature, is not simply a means of producing a certain sensation but a constituent part of a life that is thereby enriched.[71] The recognition of such values, I have suggested, is what distinguishes a broad from a narrow anthropocentrism, yet the language of domination is at least suggestive of a means–end relationship. Grundmann's attachment to the metaphor of domination therefore goes hand-in-hand with his insistence on regarding all values derivative from human interests as instrumental.

Of course, aesthetic appreciation need not always be passive, and it might be argued that it is mistaken to view aesthetic values as non-instrumental.[72] This, however, fails to rescue the domination metaphor, since the criticism requires only that there are *some* values in nature which have one or the other of these characteristics and are therefore likely to be neglected by a domination-oriented perspective. It is often argued that we value

[71] More fully, the idea is that a life may be better for including the appreciation of beauty, and that appreciation consists in a relation between the viewer and the aesthetic object, and not merely in a sensation that is only contingently (i.e. causally) related to the object and could be produced by other means. Cf. Raz 1984, pp. 188–9; Sober 1995, p. 245.

[72] A classical utilitarian, for example, would hold that the value of an aesthetic object consists solely in its capacity to cause valuable states of pleasure in an observer. The relation between observer and object is not constitutive of the valuable state, on this account, since a similar pleasure induced by any other means would be equally valuable. However, this view is undermined by our tendency to regard *authenticity* as important to our aesthetic experiences, whether in art or nature. One reason that may be given for valuing an original painting more highly than a copy or a fake, is that the former is more directly an embodiment of the intentions and actions of the artist than the latter. Similarly, if, as Goodin suggests (see the following note), one of the things we value in nature is its having been created by non-human processes (and not simply the subjective state of believing it to have been so created) then we will not regard recreated or 'faked' nature, however convincing, as its equivalent in value. If we judge an experience of faked art or nature, even one which convinces us at the time, to be less valuable than the experience of the real thing, then we will not be valuing the object of that experience instrumentally, as a mere cause of a valuable sensation, but as an essential component of the valuable state, taken to be a relation between us and the object.

nature for its 'otherness' or independence of humanity; for providing a (relatively) fixed background or context in which to locate ourselves and our activity.[73] Whether or not this is thought of as a form of aesthetic value, it is certainly a form of value which a broad anthropocentrist can recognise and in which a kind of passivity in relation to nature plays an essential role. Indeed it might be said that the value ascribed to nature in such formulations depends precisely on the nature in question *not* having been dominated or mastered by human agency. Valuing nature in this way will give us reason to protect at least some areas of relative wilderness, for example by limiting human intervention in designated conservation areas. This, however, is hard to reconcile with the injunction to dominate nature, except by stipulating a sense of 'domination' so far from ordinary usage as to deprive the metaphor of any useful function.

1.4 Conclusion

I have attempted in this chapter to provide a preliminary account of what ecological problems are and how they should be evaluated. On the first question we have seen that, owing to the vagueness of the nature/society contrast, a sharp distinction cannot be drawn between ecological and other social problems. The vagueness arises from the fact that much of what we call nature has been modified by human activity while the elements that make up human society and its products remain natural in important ways. This interpenetration of the natural and the human or social is a prominent feature of Marx's materialism and will be examined further in chapter 4. As far as the identification of ecological problems is concerned, I have argued that we are in practice largely able to agree on what counts as an ecological problem, and this is reflected in the widespread recognition of certain categories of ecological problem, namely pollution, depletion of resources and disruption of natural systems.

On the second question, I have argued that ecological problems should be evaluated from a perspective that I have called a broad and extended anthropocentrism; that is, a perspective which regards all values as being derivative from the interests of humans or other sentient creatures, but

---

[73] A good example is Goodin 1992, where it is argued that the value of natural objects lies in 'the history and process of their creation', and in particular 'the fact that they have a history of having been created by natural processes rather than by artificial human ones' thus providing 'a larger context', something 'outside of ourselves', in which to situate our plans and projects. Goodin characterises his theory as a 'halfway house' between shallow and deep ecology, but within the framework given here it would count as a broad anthropocentrism. For a related view proposed in a different context, see Norman 1996.

which recognises that the values so derived need not only be instrumental. This broad anthropocentrism requires that the classification of ecological problems into pollution, resource depletion and disruption of ecosystems be either qualified or extended. 'Resource' is often understood in economic terms as a source of economic wealth, but even if it is interpreted more widely than this, there is a natural reading of the term (which I provisionally adopted in section 1.2) which would limit its application to objects that are *instrumentally* useful to human beings. The adoption of a broad anthropocentric perspective therefore requires us either to introduce a further category of environmental problem to cover the destruction of natural entities whose value is non-instrumental, or to stipulate that 'resource' is understood in a wider sense which includes such values. The latter, however, would tend to underpin the view that things like wilderness and beautiful landscape have merely instrumental value, as means for producing pleasure, and to obscure the fact that notions like efficiency and substitution which apply to the use of ordinary resources may not apply to these non-instrumental goods. The loss of such goods is therefore best conceived as a distinct category of environmental problem. Such problems are, however, often a result of other environmental problems. Loss of landscape or wilderness often results from its being utilised as a resource (for example when hillsides are quarried, and forests felled for timber or cleared for agriculture), the pressure for which increases as the availability of alternative resources declines. Alternatively such losses may result from pollution (for example the damage to forests caused by acid rain) or the breakdown of natural systems (landscape change caused by global warming).

The way in which ecological problems may be understood in terms of resource depletion will be explored in the next chapter, where I will examine the influence of one particular account of resource depletion – Malthus's theory of population – upon recent environmental thought. The chapter will also make use of the idea, discussed above, of the interconnection of nature and society, by showing how the so-called 'natural limits' in fact arise out of a combination of natural, social and technological factors.

2 Marxism and the green Malthusians

In the last chapter I rejected the 'ecocentric' perspective of many green thinkers, according to which the range of our morally permissible actions is limited or constrained by values in nature that are independent of the interests of humans and other sentient creatures. In this chapter I will examine the claim that nature constrains our actions in another way: by presenting us with a set of non-moral facts which we must acknowledge and adapt to for the sake of our own well-being. This idea of natural limits as material constraints upon human action and especially upon economic growth expresses the important truth that the range of actions open to us, and the effects of those actions, depend in part upon facts about the world that are beyond our control. I will argue, however, that even when natural limits are understood in this factual sense, they remain dependent upon human evaluations, interests and activities to a greater extent than has often been recognised in environmental thought.

The claim that there are definite limits to growth to which society must adapt has been central to the development of environmental thought and remains one of its most important themes. The dramatic and shocking (though subsequently disputed) conclusions drawn from this claim in the late 1960s and early 1970s played an important part in bringing environmental issues to public prominence, and the same contention underlies the claim of green parties to possess a distinct political ideology. Indeed, despite some qualifications to the 'no-growth' slogan, many greens see the question of growth as the most fundamental political divide, in comparison with which 'the debate between the protagonists of capitalism and communism is', as Jonathon Porritt puts it, 'about as uplifting as the dialogue between Tweedledum and Tweedledee'.[1]

[1] Porritt 1985, p. 44.

The limits-to-growth theme has a Malthusian character which is only partially acknowledged by its proponents. In the first half of this chapter I will demonstrate the various ways in which recent environmental thought has made use of arguments strongly resembling those propounded by Malthus more than a century earlier. I will then examine Marx and Engels's critique of Malthus in order to show that, far from being evidence of an essentially anti-ecological bent, as is sometimes claimed, their critique contains elements which may be applied today as a valid and necessary corrective to the rather crude and rigid conceptions of natural limits that are frequently encountered in green thought.

2.1 Population and resources: Malthus's mathematical model

The general outlines of Malthus's theory of population are well known but are worth reiterating here as it will be necessary to refer to the theory in subsequent sections of this chapter. Malthus's central thesis concerns the relationship between population growth and means of subsistence:

Population, when unchecked, increases in a geometrical ratio. Subsistence increases only in an arithmetical ratio. A slight acquaintance with numbers will show the immensity of the first power in comparison of the second.[2]

But, he continues, because food is necessary for life, '[t]his implies a strong and constantly operating check on population from the difficulty of subsistence. This difficulty must fall somewhere and must necessarily be felt by a large portion of mankind.'

The claim concerning the merely arithmetical growth of means of subsistence has as its premises the finite area of available land and the limited potential for increasing its yield. It is only this, in Malthus's view, which prevents edible plants and animals from undergoing the same geometrical increase that he regards as characteristic of humans. He does not mean to imply that an arithmetical growth of food supplies is inevitable; indeed, he has no particular reason for specifying an 'arithmetical' model; he simply thinks that this is the greatest increase imaginable to even the most enthusiastic speculator, an increase which would in a few centuries make every acre of land as prolific as a garden.[3]

The 'checks' imposed upon population growth by scarcity of the means of subsistence are divided by Malthus into two categories: 'preventive

---

[2] Malthus 1986b, p. 9. Contemporary commentators tend to speak of 'exponential' and 'linear' growth where Malthus uses the terms 'geometric' and 'arithmetic'. In what follows I will use the two pairs of terms interchangeably, according to context.   [3] *Ibid.*, p. 12.

checks', disincentives to marriage leading to a lower birth rate; and 'positive checks', the increased death rate resulting from undernourishment, disease, war and other causes believed by Malthus to be consequent upon overpopulation. Because of these checks, the postulated natural tendency of population to grow geometrically can operate only under abnormal circumstances and for limited periods. Malthus supports his assertion that such a tendency nevertheless exists by appealing to the experience of 'unchecked' population growth in underpopulated communities such as the English colonies in North America.[4] However, his attempt to prove the universality of this tendency by asserting that 'the passion between the sexes is necessary, and will remain in nearly its present state' suggests that his model of geometrical growth owes as much to *a priori* reasoning as to his empirical case studies. It is true that a constant birth rate repeated in successive generations will produce a geometrical growth in population, assuming that the death rate remains constant and that the birth rate is above replacement level. However, if the birth rate is *at* or *below* replacement level then population will remain steady or decline. Malthus's studies of North America may, as he claims, reveal a birth rate above replacement level, but to make it an axiom that this rate will be constant for all times and places is to assume what is to be proved. Even leaving aside the empirical validity of Malthus's claims, the fact is that according to his own theory human population *is* checked, strongly and constantly. Why then does he give such prominence – and attribute the status of universal law – to a rather arbitrary arithmetical model and a geometric model which is at best an abstract idealisation?

Malthus adopts these models in order to illustrate and add weight to his belief that human misery, caused by scarcity of the means of subsistence, is inevitable, and that efforts to alleviate poverty and hunger by increased production or redistribution of food are futile and even counterproductive. No increase in food supply, he argues, can provide more than a temporary respite from hunger since it will rapidly be outstripped by population growth. While professing sympathy for the goals of humanitarian reformers, Malthus contends that the inequality of the powers of population and food production furnishes a conclusive argument 'against the perfectibility of the mass of mankind'.[5] Malthus also opposed even the more modest attempts to ameliorate the condition of the poor, such as provision of parish assistance under the English Poor Laws, on the grounds that such measures simply exacerbate the problems faced by the poor, by removing

---

[4] *Ibid.*, p. 8.   [5] *Ibid.*, p. 10.

one of the checks to further population growth. Under such conditions, he held, population would grow geometrically until once again it outstripped food supplies, leaving more people than before living in poverty. Instead, Malthus advocated the system of workhouses, insisting that their regime should be sufficiently harsh that they would serve as a deterrent against the raising of families by those who could not be confident of their continued ability to support them.[6]

In the first edition of his *Essay on Population* Malthus offers no escape from the consequences of his 'law'. Restriction of population growth by 'preventive checks' offers no alternative since, in his view, such checks come about only in response to a scarcity of means of subsistence that has already arisen. In addition he argues that so strong is the 'constant effort towards an increase of population' that such measures lead 'almost necessarily, though not absolutely so', to 'vice', and are therefore unable to alleviate the distress of the lower classes.[7] The latter claim needs some explanation, since one might assume Malthus's point to be that the vice resulting from abstinence from marriage will produce increased numbers of illegitimate births, thereby undermining the attempt to control population. In fact, Malthus accepts that abstinence from marriage, even when accompanied by an increase in promiscuity, may lead to a reduction in birth rate; indeed, he holds that promiscuity may itself tend to reduce the birth rate by damaging women's fertility.[8] Instead, the reference to vice appears to function as an independent moral premise, intended to show that the results of imposing preventive checks on population growth are as undesirable (since vice itself contributes to the very 'distress' of the masses that the preventive checks were intended to ameliorate) as the positive checks that will occur in their absence. On either reading, however, the conclusion of Malthus's argument is that misery is inevitably the lot of the masses.

In later formulations Malthus allows a greater significance to what he terms 'moral restraint', defined as 'abstinence from marriage, either for a time or permanently, from prudential considerations, with a strictly moral conduct towards the sex in the interval', describing this as a 'mode of keeping population on a level with the means of subsistence, which is perfectly consistent with virtue and happiness', and even allowing that it may be fostered by an appropriate government policy and social environment.[9] If Malthus is conceding here that by such measures the birth rate *can* be

---

[6] *Ibid.*, pp. 29–38; cf. Malthus 1986d, pp. 238–9.   [7] Malthus 1986b, p. 14.
[8] See Malthus 1986c, pp. 15, 20–21; Malthus 1986d, p. 203.
[9] Malthus 1986d, pp. 203–6. See also Malthus 1986c, pp. 465–73.

significantly reduced, then this further undermines his claim to have stated a universal law of population growth. He remains pessimistic, however, about the potential for moral restraint and does not allow it sufficient force to undermine his opposition to redistributive measures.

2.2 Environmentalist appropriation of Malthus

2.2.1 Malthusianism on a global scale

In order to trace the Malthusian motif in environmental thought I will first consider two writers whose engagement with environmental issues began in the late 1960s and in whose writings the influence of Malthus can be in no doubt. Paul Ehrlich's book *The Population Bomb* begins with an account of what he takes to be 'The Problem', defined under three revealing headings. First, 'Too Many People', in which he examines and speculatively extrapolates from past exponential (or geometrical) population growth, and second, 'Too Little Food'. The Malthusian contradiction between population growth and means of subsistence could hardly be stated more clearly: 'Each year food production in undeveloped countries falls a bit further behind burgeoning population growth, and people go to bed a little hungrier. While there are temporary or local reversals of this trend, it now seems inevitable that it will continue to its logical conclusion: mass starvation.'[10] It is only in the third section, 'A Dying Planet', that Ehrlich introduces the novel element. The environmental factor is added to the Malthusian formula, as an additional constraint upon the capacity to meet growing human need by increasing food supplies. 'Our problems', he says,

would be much simpler if we needed only to consider the balance between food and population. But in the long view the progressive deterioration of our environment may cause more death and misery than any conceivable food–population gap. And it is just this factor, environmental deterioration, that is almost universally ignored by those most concerned with closing the food gap.[11]

Garrett Hardin's frequently cited essay, 'The Tragedy of the Commons', appeared in the same year as the *Population Bomb*. Its argument is explicitly located within the framework of Malthus's 'law': 'Population, as Malthus said, naturally tends to grow "geometrically", or, as we would now say, exponentially. In a finite world this means that the per capita share of the world's goods must steadily decrease.'[12] Malthus, as we have seen, bases his prediction of continued population growth to the limits of subsistence

[10] Ehrlich 1968, p. 17.   [11] *Ibid.*, p. 46.
[12] Hardin 1980, p. 101 (originally published in 1968).

on what he takes to be the near-irresistibility of the human sexual drive. The implicit premise, no longer plausible in an era of widely available birth control, is that sex leads inevitably to procreation. Hardin, however, offers an alternative mechanism to explain population growth and to show why, in his view, it will inevitably outstrip resources unless coercive measures are applied. This mechanism is based upon the logic of action in a 'commons', that is a system of ownership in which certain finite resources are common property. In such a system, Hardin argues, each individual gains the full benefit from his use of the common resources, but the costs of their use are borne jointly by the community. It is therefore rational for each individual to increase his use of the commons without limit. But when every member of the community behaves in this way the result is the destruction of the commons: 'Ruin is the destination towards which all men rush, each pursuing his own best interest in a society that believes in the freedom of the commons. Freedom in a commons brings ruin to all.'[13] The logic of the commons, Hardin maintains, applies to a variety of problems, including overconsumption of various resources and pollution, but most crucial, he believes, is its effect in producing population growth. It is because population levels have risen that pollution and resource depletion have reached serious proportions, and so, he argues, the most important task facing us is to abandon the 'commons in breeding' that allows people to procreate without bearing the full costs.[14]

Both Ehrlich and Hardin, then, adopt a Malthusian explanation of human poverty and environmental degradation, seeing the cause of these phenomena in a rate of exponential population growth that inevitably outstrips finite means of subsistence and will rapidly exhaust any conceivable increase that may be procured in the latter. Like Malthus they conceive means of subsistence primarily in terms of food supply. Their adoption of Malthusian explanations leads Ehrlich and Hardin to Malthusian 'solutions': since population growth will neutralise the effects of increased food supply, the only hope is to curtail population growth, and – like Malthus – they believe that easing the conditions of the worst-off will, at least in some cases, have a detrimental effect upon this objective. Hardin, for example, in a passage reminiscent of Malthus's criticism of the Poor Laws, blames the welfare state for the growth of population:

If each human family were dependent only on its own resources; *if* the children of improvident parents starved to death; *if*, thus, overbreeding brought its own 'punishment' to the germ line – *then* there would be no public interest in controlling the

---

[13] *Ibid.*, p. 104.   [14] *Ibid.*, pp. 106, 113.

breeding of families. But our society is deeply committed to the welfare state, and hence is confronted with another aspect of the tragedy of the commons.[15]

But whereas Malthus focused his attention on the relationship between food supply and population in individual countries, many environmental problems in the late twentieth century are global in character, and Ehrlich and Hardin accordingly extend their Malthusian principles to the international arena. Both attempt to model the global environmental crisis by means of striking and emotive metaphors, whose correspondence to the real situation has, however, rightly been questioned.[16] Hardin compares the developed countries to lifeboats which would put themselves in danger of sinking if they went to the assistance of those (Third World inhabitants) whose boats are already overloaded or who are already floundering in the sea. He concludes:

If the poor countries received no food aid from outside, the rate of their growth would be periodically checked by crop failures and famines. But if they can always draw on a world food bank in time of need, their population can continue to grow unchecked. In the short term a world food bank may diminish that need, but in the long term it actually increases that need without limit.[17]

Ehrlich's analogy of a single 'Spaceship Earth' is at first sight less exclusionary than that of Hardin in that it places more emphasis on the common interests of all 'passengers'. However, in elaborating the picture, he makes it clear that preventing the craft from becoming overloaded is a priority which may require the needs of the steerage passengers to be overridden. To this end Ehrlich advocates that the first-class passengers (inhabitants of the developed countries) control their own populations, by coercion if necessary, and use their power to force such a program upon the others.[18] Elsewhere he describes a scenario in which one-fifth of the world's population dies by starvation as a result of food supplies being withheld from Third World countries, as having 'considerably more appeal' than any realistic alternative.[19]

2.2.2 Malthus broadened: 'The Limits to Growth'

A second stage in the application of Malthusian ideas to environmental issues can be discerned in the immensely influential first report to the Club

---

[15] Ibid., p. 107.
[16] For a valuable critique of Malthusian accounts of poverty and the images they employ, see O'Neill 1986, especially pp. 17–18 and 58–9.
[17] Hardin quoted in O'Neill 1986, p. 58.   [18] Ehrlich and Harriman 1971.
[19] Ehrlich 1968, p. 80.

of Rome, *The Limits to Growth.*[20] This first appeared in 1972, a few months after an account prepared by the editors of the *Ecologist* magazine, *A Blueprint for Survival.*[21] The latter, as its title suggests, concentrates more on proposed solutions to environmental problems than on their causes. It does, however, give an account of these causes which is essentially the same as that of the *Limits to Growth.*[22] These accounts share a characteristic which may appear to distance them from Malthusian forms of thought, which is that they relegate population growth from being the single or dominant cause of environmental problems to being just one of several joint causes. In reality, though, they broaden the Malthusian influence by applying to each of the supposed causes the same Malthusian logic of impending crisis brought about by the conflict between unlimited, and indeed exponential, growth in demand and finite means to satisfy that demand. Thus the *Blueprint* declares:

It should go without saying that the world cannot accommodate this continued increase in ecological demand. *Indefinite* growth of whatever type cannot be sustained by *finite* resources. This is the nub of the environmental predicament. It is still less possible to maintain indefinite *exponential* growth – and unfortunately the growth of ecological demand is proceeding exponentially. . .[23]

The *Limits to Growth* surveys five factors which are considered to be causes of environmental crisis: 'All five elements basic to the study reported here – population, food production, industrialization, pollution, and consumption of natural resources – are increasing. The amount of their increase each year follows a pattern that mathematicians call exponential growth.'[24] Its authors, like Malthus, see the combination of exponential growth in demand and finite resources as undermining the possibility of meeting human needs. But, whereas Malthus considers only the 'checks' imposed by food supply, the *Limits to Growth* also includes in its model the limited supplies of certain non-renewable resources used by industry, and limits to the capacity of the earth's natural systems to absorb waste materials: 'If the present growth trends in world population, pollution, food production, and resource depletion continue unchanged, the limits to growth on this planet will be reached sometime within the next one hundred years.

---

[20] Meadows *et al.* 1974.   [21] Goldsmith *et al.* 1972.
[22] This is not surprising since the authors of the *Blueprint* drew upon the research of the *Limits to Growth* team. See O'Riordan 1976, p. 52.   [23] Goldsmith *et al.* 1972, p. 17.
[24] Meadows *et al.* 1974, p. 25. Strictly speaking, the statement that food production is increasing exponentially contradicts Malthus's premise that food production can increase, at best, linearly. Taken as a whole however, the argument closely follows that of Malthus: whereas demand for increased food supply will *continue* to rise exponentially, the actual rise in production will sooner or later be 'checked' by physical constraints, primarily the availability of uncultivated land. See pp. 46–54.

The most probable result will be a sudden and uncontrollable decline in both population and industrial capacity.'[25]

One of the chief criticisms of the *Limits to Growth* has been that it gives insufficient weight to the fact that technological advances may improve the efficiency with which resources can be extracted and used, enable substitution of new materials and control pollution.[26] Its approach to technology mirrors Malthus's response to measures designed to increase the amount of food available for the poor. That is, that unless the growth of demand is controlled, such measures are bound to fail, any conceivable increase in supply being rapidly absorbed and overtaken by uncontrolled growth in demand: 'Such means may have some short term effect in relieving pressures caused by growth, but in the long run they do nothing to prevent the overshoot and subsequent collapse of the system.'[27] Furthermore, just as Malthus believed the Poor Law provisions to be dangerous because they inhibited recognition of the need for population control, so the authors of the *Limits to Growth* see danger in the ability of technology to 'relieve the symptoms of a problem without affecting the underlying causes': 'Faith in technology as the ultimate solution to all problems can thus divert our attention from the most fundamental problem – the problem of growth in a finite system – and prevent us from taking effective action to solve it.'[28]

Like Malthus, the authors of the *Limits to Growth* see the solution to the problem of expanding demand for finite resources not in increasing the supply of the means of subsistence, but in curtailing the growth of demand. But they are concerned with the limits not only of food supply but also of the resources used by industry, and such resources, unlike food, have no natural ceiling to their per capita consumption. Consequently the prevention of further growth in demand for these resources is perceived as entailing abolition of economic as well as population growth. Going further, the report advocates that consumption of resources be cut in order to extend the time for which the proposed steady-state economy may be maintained. In making these recommendations the *Limits to Growth* shows itself to be considerably more optimistic than Malthus himself about the possibility of suppressing growth in demand. However, because the report has little to say about how such a regime may be implemented, its proposals have been considered by many to be naive.

---

[25] *Ibid.*, p. 23.   [26] For example, Cole *et al.* 1973.   [27] Meadows *et al.* 1974, p. 157.
[28] *Ibid.*, p. 154.

2.2.3 *Malthus softened: emancipatory environmentalism*

Recent environmental thought, emanating principally from the green movement, has appeared to adopt less rigid and consequently less Malthusian views on growth. On investigation, however, it will be seen that the basic explanatory structure of Malthus's 'law' remains in place. An early example of this apparently more flexible approach is the second report to the Club of Rome, *Mankind at the Turning Point*. Its authors rightly complain of the naivety of unqualified arguments 'for' or 'against' growth. 'To grow or not to grow,' they write, 'is neither a well-defined nor a relevant question until the location, sense, and subject of growing and the growth process itself are defined.'[29] In order to provide the required definition, the authors make use of a distinction, borrowed from descriptions of biological growth processes, between *undifferentiated growth*, which is merely a quantitative and uniform multiplication of the whole, and *organic growth* or growth with differentiation, in which the amount and type of growth varies for different parts or organs, but is co-ordinated in accordance with the needs of the organism as a whole. Applying this analogy to the 'world system' enables them to advocate curbs to growth of some types and in some areas, together with more growth in others:

> While undifferentiated growth is assuming truly cancerous qualities in some parts of the world, the very existence of man is threatened daily in some other parts by lack of growth; e.g., in regional food production. *It is this pattern of unbalanced and undifferentiated growth which is at the heart of the most urgent problems facing humanity – and a path which leads to a solution is that of organic growth.*[30]

This insistence on viewing growth in qualitative and not just quantitative terms will be important later, in assessing Marx's commitments to growth; growth, that is, of human needs and of the productive forces.

A similar view is put forward by Jonathon Porritt, anxious to free the green movement from charges of dogmatic opposition to economic growth:

> It is surely self-evident that we cannot continue expanding at past rates of growth, and yet since the war we've made this one measurement the ultimate arbiter of

---

[29] Mesarovic and Pestel 1975, p. 3.

[30] *Ibid.*, p. 7. A possible objection to this account is that existing patterns of growth are not 'undifferentiated' in the sense of being uniform multiplications of the whole. Rates and types of growth do vary between different parts (whether geographical divisions or industrial sectors) of the world economy. This account therefore needs to be reformulated: the problem is not that growth is literally undifferentiated but that it is inappropriately differentiated and that the differentiation is not co-ordinated in accordance with human (or other) needs.

social progress. Because of our opposition to this manifest absurdity, ecologists are always seen as the 'no-growth party', the 'zero-growthers', though such a position is obviously just as absurd as that adopted by the 'infinite-growthers'. It is our contention that there will always continue to be *some* economic growth: in the developed countries there will be limited growth in certain sectors of the economy, even though the overall base will no longer be expanding; in the Third World, there will have to be substantial economic growth for some time, though with much greater discrimination as regards the nature and quality of that growth. All economic growth in the future must be sustainable; that is to say, it must operate within and not beyond the finite limits of the planet.[31]

Porritt's approach appears to match the views of environmental economists who have criticised GNP as a measure of development on the grounds that it fails to take proper account of the negative utility of pollution and depletion of natural resources, and indeed Porritt has endorsed their calls for changes in the way economic growth is quantified.[32] The quoted passage, however, is marred by a failure to specify what is meant by 'growth', which kinds of growth are regarded as acceptable and which are not. The goal of 'sustainable development', alluded to by Porritt and seized upon by governments anxious for environmental credibility, has also been criticised by environmentalists, for being too open to competing interpretations and hence too easily used as a smokescreen for 'business as usual'.[33] It is clear, however, that for Porritt, sustainable development means *less* economic growth and not simply growth of a different kind. The milder strictures in comparison with the *Limits to Growth* result not from a radically different theoretical framework but from a less pessimistic assessment of the rate at which the gap between growing demand and finite supply of resources is closing.[34] This may allow more time for adjustments to the economy in the North and catching-up development in the South, but does not remove the threat of Malthusian scarcity. The Malthusian principle that scarcity must be addressed not by increasing supply but by curtailing demand remains, for Porritt, a defining feature of green politics: 'If you want one simple contrast between green and conven-

[31] Porritt 1985, p. 120.

[32] GNP is discussed, for example, in Pearce *et al.* 1990. Cf. Porritt 1991, p. 15.

[33] See, for example, Ekins 1991, p. 29; Jacobs 1997, pp. 3–5. The coining of the term 'sustainable development' is usually attributed to the Brundtland Report (World Commission on Environment and Development 1987). Since this chapter was first drafted the United Nations Conference On Environment and Development (the Rio 'Earth Summit') in 1992 has provided further evidence of governmental enthusiasm for the rhetoric of sustainable development; however, the record of putting into practice promises made at Rio, together with the equal enthusiasm for achieving a GATT agreement on free trade (promoted as a means of enhancing economic growth, but with no corresponding attention to how that growth may be directed), has done nothing to ease doubts about its sincerity.

[34] Porritt 1985, p. 26.

tional politics, it is our belief that quantitative demand must be *reduced*, not expanded.'[35]

Another apparent step away from Malthusianism in recent environmental thought is a reduced emphasis on the issue of population growth. One reason for this may be the perceived unpopularity of population restraint in the context of the green movement's entry into electoral and pressure-group politics.[36] Reticence on this issue does not, however, imply a fundamental change of policy. Porritt expresses the view of many greens in rejecting coercive restrictions on population as 'an unacceptable and morally repugnant infringement of human rights' but nevertheless insisting that 'the costs of procrastination rather than action are appalling to contemplate, and for many will mean the difference between survival or extinction'.[37] The rejection of coercive measures is advanced as the policy calculated to be most effective in achieving the goal of curtailing population growth. Not all advocates of green politics, however, adopt such an attenuated Malthusianism as Porritt. Irvine and Ponton, for example, praise Malthus and Ehrlich for their frankness in addressing the 'crisis in overpopulation', and advocate a 'stick and carrot' approach to its solution. Like Ehrlich and Hardin they are prepared to use Malthusian policies on an international scale as a lever to enforce population control: 'In terms of foreign aid, the cruel truth is that help given to regimes opposed to population policies is counterproductive and should cease. They are the true enemies of life and do not merit support.' Although such policies may seem Draconian, they write, 'they are mild compared to what will be required if active steps are not taken now'.[38]

What we have seen is that despite a tendency to move away from harsh Malthusian policy recommendations, the concept of limits to growth, premised upon the Malthusian idea of scarcity resulting from the clash between an unlimited (and exponential) growth in demand and absolute

---

[35] *Ibid.*, p. 136.

[36] It might be objected that population restraint applied to developing countries would not be unpopular among Western electorates. However, there may still be reason for green parties to play down population issues. Firstly, if a policy of population restraint were adopted it would only contingently be possible to exclude Western populations. Population growth may be lower in the developed countries, but in many cases population density is higher, so any upward turn in population trends in the developed countries would bring them within the scope of any consistently applied coercive policy. Secondly, parties which adopted a coercive policy with respect to developing countries would be politically vulnerable to the charge (whether justified or not) that this represents the thin end of the wedge and that such parties are prepared more generally to remove people's freedoms in pursuit of environmental objectives. Thirdly, green parties may well have their greatest potential appeal among sections of the electorate who *would* object to coercive policies applied to Third World populations.

[37] Porritt 1985, p. 193. Cf. Dobson 1990, p. 95.   [38] Irvine and Ponton 1988, p. 23.

limits to the earth's ability to supply this demand, remains a central pillar of environmental thought. This conception of scarcity has been broadened from Malthus's own concern with population-led growth in demand for food, to a more general view of scarcity of the resources required for the satisfaction of human needs, growth in demand for which is determined by per capita consumption as well as population. A second shift has been the adoption of a more flexible attitude to the possibility and permissibility of limited continued growth (limited by type and location as well as quantity). This flexibility depends upon a less pessimistic assessment of resource scarcity, but has not led to an abandonment of the underlying logic, a point which is expressed by Dobson when he notes that one of the chief criticisms of the *Limits to Growth* has been that its pessimistic predictions of resource exhaustion have proved wildly inaccurate, but responds that greens 'have learned to accept the detail of these criticisms while continuing to subscribe to the general principles of the limits to growth thesis'.[39]

2.3 Marx and Engels on Malthus

The remainder of this chapter is devoted to an analysis of Marx and Engels's critique of Malthus, giving particular attention to the Malthusian forms of argument that I have shown to be characteristic of environmental thought. I use the term 'critique' in the singular because nothing in the material that I examine indicates that the attitudes of Marx and Engels to Malthus differ in any important respect.

Two writers on this subject have disputed the Malthusian characterisation of contemporary environmentalism. Ted Benton and K. J. Walker both accept Marx and Engels's arguments against Malthus as correct in their own terms, but argue that the content and context of modern environmentalism are significantly different from those of Malthus's theory, such that 'the classical Marxian critique of Malthus will not serve our purposes today.'[40] Benton argues that Marx and Engels overreacted to Malthus in such a way as to deny the existence, or underestimate the significance, of all natural limits, and that it is therefore necessary to revise Marxism in order to create a green historical materialism. Walker, on the other hand, believes that Marx and Engels did recognise natural limits but that their followers have ignored this aspect of their thought, misinterpreting their critique of Malthus, in order to treat environmental problems simply as

[39] Dobson 1990, p. 80.   [40] Benton 1991, p. 247. Cf. Walker 1979, pp. 29–31.

consequences of capitalist relations of production. I will argue that Marx and Engels's critique of Malthus does not imply a denial of environmental limits and that it is relevant to contemporary environmental problems, giving an insight into the weaknesses of the approaches I have been examining.[41]

2.3.1 Malthus's ideological agenda

The most visible and at the same time the least constructive element of Marx and Engels's critique of Malthus is their attack on what they take to be the conservative ideological agenda accompanying his theory. In several works of the 1840s, including Engels's *The Condition of the Working Class in England in 1844*, Malthus is attacked for justifying the workhouse, disguising the real causes of poverty and treating poverty as 'the eternal destiny of mankind'.[42] Marx, in his later works, portrays Malthus's theory as a part of the establishment reaction to the French Revolution and an 'antidote' to the progressive ideas promoted by it.[43]

Several distinct claims can be discerned here. One is simply that Malthus's theory has conservative political implications, implying as it does that attempts to relieve poverty must inevitably fail. As Marx puts it, Malthus's theory, if true, would prove 'that socialism cannot abolish poverty, *which has its basis in nature*, but can only make it *general*, distribute it simultaneously over the whole surface of society!'[44] A second claim is that Malthus's conservatism is not simply a consequence of his scientific views, but rather that his political conservatism provides the motivation for views on population that are presented as scientific but in fact lack scientific justification. Thus Marx describes Malthus's theory as a 'lampoon', 'apologia' and 'panegyric', and claims that Malthus 'only draws such conclusions from the given scientific premises . . . as will be *"agreeable"* (useful) to the aristocracy against the bourgeoisie and to both *against* the proletariat'. Marx also cites the conservative implications of Malthus's theory to explain its widespread acceptance by the English ruling classes, and he

---

[41] Benton and Walker are criticised in Grundmann 1991a, p. 51 for concentrating on 'one of the two main ecological problems', scarcity of resources, and 'almost completely neglecting the other: pollution'. But pollution as well as resource depletion may be considered within the conceptual framework of natural limits, as we shall see, and Benton does in fact refer to pollution in this context, as he points out in reply to Grundmann (Benton 1992, pp. 56–7).

[42] *The Condition of the Working Class*, p. 570. See also Marx's 'Critical Marginal Notes' and Engels's *Outlines of a Critique of Political Economy*.   [43] *Capital*, vol. I, p. 766.

[44] *Critique of the Gotha Programme*, p. 324.

and Engels further pour scorn on Malthus's scientific credentials by frequent accusations of plagiarism.[45]

None of this, however, implies that Malthus's theory is false. The incompatibility of Malthus's theory with emancipatory politics may explain Marx and Engels's hostility to Malthus (and by extension may explain the suspicion with which environmentalism, with its Malthusian associations, has often been regarded by the contemporary political Left[46]), but to reject his theory on this basis would be to engage in just the sort of ideologically driven theorising for which Marx and Engels criticise Malthus. The remaining claims, about the motivation and conduct of Malthus and his supporters, even if true, constitute only circumstantial evidence against the theory itself. What is needed in order to diffuse the Malthusian and (neo-Malthusian) challenge to Marxism and to progressive politics generally is not an attack on Malthus's motivation, but a reasoned critique of his claim that demand for resources inevitably runs up against natural limits, rendering scarcity unavoidable.

2.3.2 The critique of Malthus on population

Within Marx and Engels's critique of Malthus can be found challenges to both of his key claims: that population tends to grow exponentially and that food supply can at best undergo a linear increase, limited by the availability of fertile land. In his *Outlines of a Critique of Political Economy*, Engels argues that population growth does not inevitably follow a geometrical progression but is influenced by social conditions and, under appropriate conditions, may be subjected to conscious control – something which, as we have seen, Malthus partly concedes in later versions of his theory by allowing that 'moral restraint' may have some efficacy and may be promoted by appropriate government policy. But while Malthus downplays the role of government in promoting moral restraint, limiting it chiefly to the protection of private property and civil liberties,[47] Engels draws upon Malthus's concession to argue the need for social transformation:

We derive from it the most powerful economic arguments for a social transformation. For even if Malthus were completely right, this transformation would have to be undertaken straight away; for only this transformation, only the education of the masses which it provides, makes possible that moral restraint of the propaga-

---

[45] *Theories of Surplus Value*, part II, pp. 117–19 and part III, p. 61. See also part II, pp. 114–17 and 120.

[46] See, for example, Enzensberger 1974, discussed in the last chapter; also Benton 1989, p. 52 and Benton 1991, p. 252.   [47] Malthus 1986d, pp. 205–6.

tive instinct which Malthus himself presents as the most effective and easiest remedy for over-population.[48]

Malthus's 'geometric' model of population growth is also rejected by Marx. Like Engels, he maintains that human population growth is a function of social and historical as well as natural factors. He rejects Malthus's view that 'the *increase of humanity* is a purely natural process, which requires *external restraints*, *checks*, to prevent it from proceeding in geometrical progression',[49] and he cites data purporting to show that population does not always rise to the limits of subsistence but has sometimes lagged behind the growth of production or even declined while production has increased.[50]

Whatever the strengths of Marx and Engels's reasons for rejecting Malthus's geometrical model of population growth, the general tenor of their conclusions today seems relatively uncontroversial. Their view of population growth as socially conditioned is, as Benton points out, closer to the conventional wisdom of contemporary demographers than Malthus's 'quasi-naturalistic' model.[51] In particular, the theory of

---

[48] *Outlines of a Critique of Political Economy*, p. 439.

[49] *Grundrisse*, p. 606. While Marx is undoubtedly right to argue that human population growth is not a purely natural process, and that it can fail to follow a geometrical progression for reasons other than brute scarcity or its threat, it should be noted that Marx's political hostility to Malthus and his penchant for a neat dialectical inversion led him to overstate his case. According to Marx, Malthus has reversed the true situation by presenting human population growth as a geometrical process subject only to external checks, while the growth of crops and other plants possesses its own internal or immanent limits. Against the latter claim Marx writes that, without external checks, 'The ferns would cover the entire earth. Their reproduction would stop only where space for them ceased. They would obey no arithmetic proportion. It is hard to say where Malthus has discovered that the reproduction of voluntary natural products would stop for intrinsic reasons, without *external checks*. He transforms the immanent, historically changing limits of the human reproduction process into *outer barriers*; and the *outer barriers* to natural reproduction into *immanent limits* or *natural laws* of reproduction.' (*Ibid.*, p. 607.)
Malthus does indeed assert that (with the partial exception of 'moral restraint') the checks on human population are external, but he does not hold the growth of means of subsistence to be subject to internal or intrinsic limits. As my earlier exposition of his arithmetical model makes clear, this model is a somewhat arbitrary attempt to express the net result of the tendency towards growth of the means of subsistence together with checks imposed by scarcity of land. There is no suggestion that the intrinsic tendency towards population growth is different in the case of plants, animals or humans. Malthus's reason for combining growth-tendency and external checks within an arithmetical model for means of subsistence but not for humans is that in the former case it is only the net result that we are interested in (i.e. the amount of food available for consumption) whereas in the case of humans we are interested in the unchecked growth-tendency and its supposed mismatch with the net growth of means of subsistence.

[50] 'The War Question. – British Population and Trade Returns. – Doings of Parliament', p. 247; 'Manufactures and Commerce', p. 493. Note, however, that since Marx's figures include the effects of emigration, they do not disprove any Malthusian claim about the pattern of *global* population growth.   [51] Benton 1991, p. 257.

demographic transition contradicts Malthus by identifying a tendency of birth rates to fall once a certain degree of affluence is attained. The typical view of environmentalists from the *Limits to Growth* onwards has been that population does not *inevitably* grow until it has exhausted the means of subsistence but that it can and must be socially controlled. The point of controversy in environmental debates is not so much the *possibility* of changing the pattern of population growth as its *necessity* and *urgency*, and the kinds of policy by which it can be achieved. This – like the question of demand for natural resources generally and its other presumed determinant, economic growth – depends upon the attitude taken to the second premise of Malthusian scarcity: limits to the supply of human needs.

2.3.3 The critique of Malthus on subsistence

The quotation from Porritt at the beginning of this chapter illustrates the widespread perception among environmentalists that the development of the productive forces postulated by historical materialism is a form of open-ended economic growth, comparable to that pursued by capitalism and equally incompatible with respect for environmental limits. Benton admits that Marx and Engels's arguments are decisive against Malthus, but claims that their concept of communism which 'presupposes the prior historical development of human productive powers', together with their 'experience of the dynamism of modern industrial production', led them (with some ambiguity) to reject not only the fallacious account of natural limits given by Malthus but all natural limits.[52] He maintains that because of this 'idealist, utopian moment' in their thought, and because the natural limits associated with today's environmental problems are closer and therefore more politically urgent than those upon which Malthus based his theory, the Marxist critique of Malthus is rendered inapplicable to the neo-Malthusianism of the contemporary environmental movement.

If Marx and Engels really did deny that natural limits place any restrictions on the growth of population and resource consumption, this would of course be untenable, and in conflict with their claim to offer a materialist theory which treats humans as embodied beings with physical needs. Examination of what they said about Malthus's second premise, the finitude of means of subsistence, however, does not warrant such an interpretation.

In the *Outlines of a Critique of Political Economy*, Engels argues that

[52] *Ibid.*, p. 259.

Malthus is wrong in his account of the limits of agricultural productivity. Benton and Walker draw attention to certain of his phrases which appear to demonstrate an overoptimistic assessment of the potential for increasing agricultural output. Engels asserts that progress in science is 'just as limitless and at least as rapid as that of population', and that it 'advances in proportion to the body of knowledge passed down to it by the previous generation' just as population (according to Malthus) increases in proportion to the size of the previous generation. He poses the rhetorical question: 'what is impossible to science?' And he declares that, by drawing attention to the productive power of the earth and of mankind, Malthus has made us 'for ever secure against the fear of over-population.'[53]

Two responses can be made to the charge that Engels overrated the potential of science. First, the case for regarding Engels's view of science as overoptimistic is not as clear as it might appear when contrasted with the pessimism of many environmentalists about the potential for scientific and technological responses to environmental problems. Engels's statement that scientific progress is geometrical is not so much unfounded, as Walker claims, as an unquantifiable statement that can only be regarded as rhetoric.[54] Benton and Walker both suggest that Marx and Engels were led to overestimate the powers of science by their 'experience of the dynamism of modern industrial production: something of which Malthus himself could have had but little comprehension'.[55] However, the advances in science and technology *since* Marx and Engels have been no less spectacular, and probably beyond anything that *they* could have imagined, leading Grundmann to conclude 'that Marx's view of technological dynamism corresponds more closely to reality than Benton's'.[56]

What matters, however, is not simply the rate of technological innovation, but its effects. The technological scepticism of Malthus's green successors is, as we have seen, directed not against the possibility of technological innovation *per se*, but against its potential for resolving environmental problems rather than just replacing one problem (or natural

---

[53] *Outlines of a Critique of Political Economy* pp. 439–40.

[54] This conclusion is reinforced by the fact that the Malthusian model of population growth with which Engels compares the growth of science is one which he himself rejects. Engels's purpose in making this comparison, I would suggest, was to show that optimistic views about the development of science are *no less* well founded than a Malthusian pessimism about the growth of population.

[55] Benton 1991, p. 259. Cf. Walker 1979, p. 34. A similar view is expressed by Michael Redclift (1989, p. 177) when he writes that Marx and Engels were 'imbued . . . with the Promethean spirit of the late nineteenth century'.

[56] Grundmann 1991b, p. 52. Grundmann cites in particular the examples of electronic and biochemical technologies.

limit) with another.[57] It follows that Engels might have been right about the dynamism of technology and yet have overestimated its potential for overcoming Malthusian scarcity by underestimating its harmful side-effects, but here too the evidence is inconclusive. At the time of writing his *Outlines* Engels may well have underestimated the harm that can arise as an unintended consequence of technological advance, but in later writings he displays a keen awareness of this problem, describing some of the disasters that have occurred as a result of human transformations of nature, and warning that each victory over nature 'in the first place brings about the results we expected, but in the second and third places . . . has quite different and unforeseen effects which only too often cancel the first'.[58]

The second, and more important response to the charge of technological overoptimism is that even if Engels's comments do imply an overoptimistic view of the potential of science and technology, such a view is unnecessary and peripheral to the main line of his critique of Malthus. Since Marx and Engels reject Malthus's view that population inevitably grows until checked by scarcity of food, they do not require an *indefinite* expansion of production in order to satisfy rising demand. The crux of Engels's critique of Malthus is that the limits of production to which the latter refers are far from having been reached:

[57] For example Ehrlich, in one of the passages quoted above (text to note 11) suggests that it is the environmental effects of increased agricultural productivity that will ultimately bring about the Malthusian crisis. *The Limits to Growth* (Meadows *et al.* 1974, pp. 129–33) considers a scenario in which technological innovations are able to overcome resource scarcity, and concludes growth would still be halted, not in this case by resource shortages but by pollution which cannot be totally controlled by technological means. Grundmann partly acknowledges that technological dynamism is not in itself sufficient to overcome ecological problems, but confuses the issue by assuming that the possible undesirable effects of technological innovation have a bearing only on the evaluative question of whether it is desirable or right for societies to overcome natural limits, and not on the factual question of whether it is possible for them to do so. The distinction is a specious one, since the side-effects of technology may impose limits on production that are no less real than those that the technology was intended to overcome, and since (as will be argued later) it is a general feature of 'natural limits' that their specification involves an evaluative element in determining the point at which the costs of continued growth outweigh its benefits.

[58] 'The Part Played by Labour in the Transition from Ape to Man', pp. 361–2. Even here Engels may be accused of excessive optimism, since he assumes that increasing knowledge of nature's laws will enable us to avoid such disasters, and declares that 'after the mighty advances made by the natural sciences in the present century, we are more than ever in a position to realise and hence to control even the more remote natural consequences of at least our day-to day production activities'. Engels does not claim, however, that science alone is able to free productive activity from its deleterious side-effects. Indeed, he specifically rejects such a view, arguing that the regulation of such effects 'requires something more than mere knowledge. It requires a complete revolution in our hitherto existing mode of production, and simultaneously a revolution in our whole contemporary social order.' (p. 363.)

it is absurd to talk of over-population so long as 'there is enough waste land in the valley of the Mississippi for the whole population of Europe to be transported there'; so long as no more than one-third of the earth can be considered cultivated, and so long as the production of this third itself can be raised sixfold and more by the application of improvements already known.[59]

According to this account, the overcoming of scarcity in the short term does not depend on developments in science yet to be made, but upon the rational use of resources and techniques already available. Engels's argument is not that there are no limits to the population that can be supported by the earth's fertile land, but that the limits have not been reached and therefore cannot be held responsible for the existence of poverty and hunger. The function of future scientific development in this picture is not to abolish natural limits altogether but to push them back, extending the potential for satisfaction of human needs. The fact that this process itself has limits is acknowledged by Engels in a letter to Kautsky, in which he entertains the possibility that at some future time population may have to be consciously regulated:

There is of course, the abstract possibility that the number of people will become so great that limits will have to be set to their increase. But if at some stage communist society finds itself obliged to regulate the production of human beings, just as it has already come to regulate the production of things, it will be precisely this society, and this society alone, which can carry this out without difficulty. . .[60]

The view that natural limits to the ability of agriculture to meet demand had not (in the nineteenth century) been exceeded by population is implicit also in Marx's explanation of what he took to be the real causes of poverty and starvation, misinterpreted by Malthus as overpopulation. According to Marx's theory of relative surplus population the present-day causes of these phenomena are not natural but social:

[Malthus] stupidly relates a specific quantity of people to a specific quantity of necessaries. Ricardo immediately and correctly confronted him with the fact that the quantity of grain available is completely irrelevant to the worker if he has no *employment*; that it is therefore the means of employment and not of subsistence which put him into the category of surplus population.[61]

Expansion of capital, according to Marx, requires a discrepancy between population and means of employment – 'a part of the population which is unemployed (at least relatively)' – in order to guarantee the ready

---

[59] *Outlines of a Critique of Political Economy*, p. 440.
[60] Quoted in Walker 1979, p. 28. Presumably the reason Engels describes this as an *abstract* possibility is that he anticipates that the population will have stabilised itself for other reasons before this point is reached.   [61] *Grundrisse*, p. 607.

availability of the labour needed to put surplus capital to work.[62] Relative surplus population is consequently a law of capital, though not a transhistorical law of society. One of Malthus's errors, Marx argues, is to regard '*overpopulation* as being *of the same kind* in all the different historical phases of economic development'.[63] The numbers which constitute overpopulation and the form that it takes are different under every mode of production. Overpopulation is 'a historically determined relation, in no way determined by abstract numbers or by the absolute limit of the productivity of the necessaries of life, but by limits posited rather by *specific conditions of production*'.[64] In a letter to Engels, Marx emphasises the need for reform of agriculture in order to remove the limits to production and distribution imposed by existing property relations and allow fuller realisation of its potential. 'Without that', he writes, 'Father Malthus will turn out to be right.'[65]

In summary then, Marx and Engels's critique of Malthus consists of (i) an assault on the ideological character and motivation of his theory, (ii) a denial of the claim that human needs, led by population growth, will inexorably rise until checked by scarcity of resources, and (iii) a denial of Malthus's belief that people's visible lack of the means to satisfy their needs is a consequence of immediate and absolute natural limits to the means of subsistence. This third part of Marx and Engels's critique includes the following claims: (a) natural limits were (at the time of writing) relatively remote; (b) the limits in question are not a function of nature alone but of the way in which humans interact with nature and hence are absolute only in the abstract and in practice are relative to human knowledge and social organisation which in varying degrees fetters or enables the realisation of natural potentialities; (c) human want is socially caused also in that it arises from failures of distribution and utilisation as well as inadequate production.

There is one element in this controversy that I have so far omitted to address. I have argued that because Marx rejects Malthus's account of population growth, his commitment to the satisfaction of human needs does not entail an indefinite (and exponential) growth in production. But what about the greens' attack on economic growth, or growth of per capita consumption, as another determinant of environmental impact? In order to assess the significance of this it will be necessary to consider Marx's own concept of 'growth' in more detail than can be given in this chapter. We have already seen suggestions (in the discussion above of 'differentiated

---

[62] *Ibid.*, p. 610.   [63] *Ibid.*, p. 605.   [64] *Ibid.*, p. 606.
[65] 'Marx to Engels in Manchester. London, 14 August 1851', p. 425.

growth' and 'sustainable development') that the relation between economic growth and environmental problems is not a simple one, and the matter is further complicated by controversy over the way in which concepts such as growth, progress and development are used by Marx, in relation to both the productive forces and human needs. I will therefore return to these issues in chapters 5 and 6. But before we can assess the extent to which environmental limits may constrain or thwart the developments anticipated by Marx we must establish the nature of those limits. It is to this end that I have been examining Marx's response to the Malthusian conception of environmental limits. The question we must now ask is whether Marx and Engels's critique of Malthus is applicable to the environmental problems faced today.

2.4 Marx, Malthus and contemporary environmental problems

It has been persuasively argued in empirical studies that, even today, the fact that millions go hungry is attributable not to an absolute limit to food production having been reached, but to social conditions. Sufficient food to feed everyone is, or could be, produced. People go hungry because they cannot pay for food; and consequently (as Marx put it in the passage quoted above)[66] to them 'the quantity of grain available is completely irrelevant'. Because the poor cannot pay, land that could produce food is put out of production and food that has been produced is dumped or left to rot. The buying power of the affluent creates incentives to maximise the monetary value of what is produced rather than its quantity and nutritional quality.[67] If this account is accepted we may conclude that so far as the relationship between population and food production is concerned, Marx's critique of Malthus retains its force.

Environmental problems today, however, concern a wider spectrum of needs and resources than those involved in food production. Walker warns of the danger of assuming 'that modern "neo-Malthusians" base their conclusions on evidence as poor as that available to Malthus' and argues that it is therefore 'undesirable to describe ecological limits as "Malthusian"'.[68] The real question though is not whether belief in the existence of environmental limits justifies the 'Malthusian' label, but whether Marx's criticisms of the concept of limits used in Malthus's account of food production are valid against environmental problems in general.

[66] See text to note 61 above.
[67] George 1977. See also Sen 1982, especially his conclusions in ch. 10, and O'Neill 1986.
[68] Walker 1979, p. 29.

It was noted earlier that recent environmentalism has had to acknowledge the overpessimistic nature of earlier accounts of the proximity of environmental limits. Previous warnings of imminent exhaustion of mineral reserves have been confounded by substitution of alternative resources and more efficient use. These cases, at least, correspond to Marx and Engels's view of natural limits as relatively distant and of existing scarcity as, at least in part, a social phenomenon.[69] One limit that does continue to be widely perceived as an immediate threat is that of the ability of natural systems to absorb pollution without disruption of their ability to support human life.[70] Scientific opinion is divided on the extent of the infamous 'greenhouse effect', but wherever the truth lies, the fact remains that the logic of the pushing back of mineral reserve limits applies also to this phenomenon. In the case of mineral resources the limit that matters for human societies is not the physical quantity of the mineral in the Earth's crust but the ability to manufacture objects using that mineral; this introduces mediating factors such as the degree of wastage in the use of the material, the extraction technique determining the proportion of the physically present material that is available for use, and substitution of alternative materials. Similarly, the greenhouse effect *directly* determines only a limit to emissions of carbon dioxide and other greenhouse gases (and this only once the level of 'acceptable' impact has been determined). Any limit to the production and use of energy (for example) is dependent also upon such mediating factors as the efficiency of available generating technology, availability of alternative energy sources (e.g. nuclear fusion and renewable sources), and techniques of waste disposal. The role of these mediating factors in relativising the effects of natural limits is identical to the role of those mediating factors which Marx and Engels highlighted in response to Malthus: that it is food production and not land that determines the maximum population (or 'carrying capacity') of an area, and that over the centuries changes in agricultural technique have allowed more food to be produced on a given area of land, and more land to be brought into cultivation.

Marx and Engels are by no means the only writers to have pointed out the relative character of 'natural' limits. A more systematic account, developed with the explicit intention of providing a conceptual framework for dealing with the environmental problems of the late twentieth century, is

---

[69] Consider, for example, the oil crisis of the early seventies, which was seen by some as the first indicator of an epoch of resource exhaustion. This is discussed by Hall 1974.

[70] I do not wish to imply that this is the *only* 'natural limit' that is perceived as an immediate threat; I have picked it out as an example because it is the most prominent.

offered by W. H. Matthews. Matthews starts from the observation that there are two basic determinants of what he calls 'outer limits', the second of which is ignored by the widespread view of such limits 'as very definite limits – set firmly by nature and theoretically determinable by man'.[71] The two determinants are:

(i) 'the quantity of existing resources and the laws of nature', and
(ii) 'the way man conducts his activities with respect to this natural situation'.

In the case of non-renewable resources, although the physical quantity of a given resource is fixed, 'the practical limits of exploiting that potential are a function of man's activities', the relevant factors being scientific knowledge, technical capabilities and economic factors, all of which in turn depend upon social and political choices and priorities.

Renewable resources and the environmental systems that generate them are limited not in total quantity but in the rate of extraction that is consistent with their continued reproduction. Matthews illustrates the effect of rate of extraction and its distribution over time with an analogy. In answer to the question: 'how many stairs can a person climb before he dies (reaches his "outer limits") . . . ?', he replies:

The person's 'outer limit' of stair climbing is not only a function of the state of his body but it is also a function of how he climbs the stairs. If he runs up them as fast as possible until he dies he can probably climb far fewer stairs than if he walks up them at a very slow pace. Furthermore, if he climbs stairs the way most of us do – only whenever he must in the course of his daily life – the question is almost totally irrelevant, for while he may climb millions of stairs during his lifetime he is not likely to die on a staircase or as a direct result of all that activity.[72]

The fact that natural limits are manipulable in these ways introduces social choices and priorities into the determination of where such limits lie. The process of extending a limit will involve costs which must be weighed against the costs of reducing resource use or of exceeding the limit. Matthews conceptualises this by observing that any outer limit may be defined only within a given frame of reference or 'context'. Matthews's definition of this concept is a little sketchy, but may most usefully be interpreted in the following way. The context involves two elements: the resource base under consideration and the range of costs that are considered significant. The significance of some costs, Matthews suggests, may be derived directly from the resource base under consideration.[73] In other

---

[71] Matthews 1976b, pp. 15–17.   [72] *Ibid.*, p. 18.   [73] *Ibid.*, p. 25.

cases the costs to be considered may be conceptually independent of the resource system. Matthews points to situations where the transgression of a certain limit for a resource would have implications for a country's independence; another example occurs where limits are placed on the levels of certain pollutants in order to restrict related disease to an 'acceptable' level. Thus contextualised, natural limits may be seen to vary in their importance. Those which concern the continued integrity of natural systems essential for human life are imperative for societies generally; others may be imperative for a particular social form.[74] The choice of contexts will be related to the structure of the agent's interests and needs. As an example of this Matthews observes that a country may well give higher priority to a national limit with implications for its independence than to the transgression of some global limit that the rest of the world considers very significant.[75]

Some writers have argued that, notwithstanding the effects of social and technological factors, the existence of *absolute* natural limits is entailed by the Second Law of Thermodynamics (the Entropy Law). The Second Law states that all physical processes result in the irreversible transformation of energy from ordered (low entropy) to more dispersed (higher entropy) states. Consequently, it is argued, within a closed system the reserves of usable energy and raw materials at low entropy are inevitably and irretrievably depleted; all we can do is to slow the process by reducing our consumption, perfect recycling being impossible.

Formally, such an argument is irrefutable, the Second Law being a central and undisputed plank of modern physics. Moreover, it is not in dispute that the Second Law is of importance in explaining ecological processes. However, the entropy-based argument for absolute natural limits is of little relevance to the study of contemporary environmental problems. The development of nuclear science and its associated technologies has

---

[74] An example of the latter is proposed by Gorz 1980, p. 55: the extent of air and water pollution in the Rhine valley, he argues, is reaching the point where the cost of reproducing 'the resources which were previously considered part of nature and therefore free' threatens to undermine the capitalist goal of 'producing the maximum exchange value for the least monetary cost'.

[75] It is only paradoxical to speak of the transgression of a global outer limit if the context in which the limit is defined is the total physical exhaustion of some resource. Where the limit is defined in terms of globally significant costs other than physical exhaustion of the resource, examples of transgression abound. Two such examples are the persistence of nuclear testing by a minority of nations which is believed to release radioactive material in excess of levels safe for human health, and the refusal of others to cut carbon dioxide emissions in order to prevent damaging climate change. Cases of this type also occur regularly with regard to world stocks of fish, whales, etc. where individual nations exceed what are widely agreed to be the limits of sustainable use.

shown that the quantity of energy contained in planetary matter and *theoretically* available for use is enormous. All matter contains low entropy energy in concentrations many times greater than that extracted from fossil fuels in conventional power stations; nuclear fusion offers, in principle, a technology to extract this energy from the water of the earth's oceans. As one writer puts it: 'Nature, matter in motion, is nothing but energy.'[76] The energy *actually* available for use is of course far less than this, but the limit is imposed not by the simple absolutes of the laws of thermodynamics but by the complex of difficulties associated with its extraction. Moreover, the earth is *not* a closed system. Consequently, proponents of the entropy argument are able to advocate the harnessing of solar energy as 'the greatest possible breakthrough for man's entropic problem' because processes utilising such energy 'would still consume low entropy, but not from the rapidly exhaustible stock of our globe'.[77] The *absolute* limit derived from the Second Law, therefore, can refer only to exhaustion of low entropy energy in the universe as a whole.[78] This phenomenon, known as 'heat death', was known to Engels, and depicted by him in his *Dialectics of Nature*:

Millions of years may elapse, hundreds of thousands of generations be born and die, but inexorably the time will come when the declining warmth of the sun will no longer suffice to melt the ice thrusting itself forward from the poles; when the human race crowding more and more about the equator, will finally no longer find even there enough heat for life; when gradually even the last trace of organic life will vanish; and the earth, an extinct frozen globe like the moon, will circle in deepest darkness and in an ever narrower orbit about the equally extinct sun, and at last fall into it. Other planets will have preceded it, others will follow it; instead of the bright, warm solar system with its harmonious arrangement of members, only a cold, dead sphere will still pursue its lonely path through universal space. And what will happen to our solar system will happen sooner or later to all the other innumerable island universes, even to those the light of which will never reach the earth while there is a living human eye to receive it.[79]

This is indeed an absolute natural limit, but portrayed here by Engels in its proper, distant, perspective.

2.5 Conclusion

What we have seen in this chapter is that the concept of ecological limits must be qualified in ways that are not acknowledged by the Malthusian

[76] Hall 1974, p. 196.   [77] Georgescu-Roegen 1980, p. 59. See also Rifkin and Howard 1980.
[78] Or at least the solar system, if we ignore science fiction proposals for escaping natural limits by means of interstellar travel.   [79] *Dialectics of Nature*, p. 20.

account that dominates much ecological discourse. An adequate account must acknowledge that the constraints imposed upon humans by non-human nature can only be determined in the context of certain goals or values, and relative to the state of technology, and that these in turn may be conditioned by the prevailing forms of social organisation.

Such an understanding partially blurs the distinction between *ecological* limits and other constraints upon human productive activity. After all, the production process itself can be characterised (as it is by Marx) as an inter-action between humans and nature, with technology as the means by which humans exert their influence on nature. And there is a sense in which all activity is directly constrained by nature: it is necessarily con-ducted within, and by means of, the laws of nature. So how can we draw a distinction between those limits which we would wish to call technolog-ical (for example, limits to our ability to fully automate factory production) or social (for example, Marx's concept of the 'fetters' placed by relations of production upon the development of productive forces) and those which we call ecological?

Ecological limits can be understood as resulting from situations in which the (actual or approaching) exhaustion of natural reserves or capacities has a detrimental impact upon the results of human activities. Once we char-acterise them in this way we can see that as a *general* representation the image of ecological limits as immovable barriers is inappropriate.[80] In general they are better modelled on a law of diminishing returns accord-ing to which similar incremental increases in output begin to require pro-portionately greater inputs, resulting from the growing costs of producing efficiency savings in the use of scarce resources or replacing them with sub-stitutes, and in countering the harm that our activities do to the natural systems upon which we depend. The limit itself can then be defined as the point at which the costs of increased production start to exceed the benefits, remembering that the nature and extent of the costs will depend upon the technology available and that the location of the point at which costs start to exceed benefits will depend upon the relative weights given to the various kinds of cost and benefit. Such a model should not be seen as ruling out extreme cases which approximate to the Malthusian model, where changes in productive technique offer little potential for altering the limit – either because of the inadequacy of our knowledge or because the

---

[80] As Matthews (1976b, p. 17) puts it: 'The limits are not, in all cases – nor even in most cases – explicit, predictable, discrete thresholds which if exceeded produce catastrophic results regardless of how they are approached. The mental image should not be that of the edge of a cliff where one additional step plunges one to the depths below.'

costs are resistant to technological manipulation.[81] It would be wrong, however, to treat these instances as prototypical and to adopt a general concept that would force all ecological limits into this mould.

The significance of Marx and Engels's critique of Malthus is that it involves an explicit recognition of the roles that technology and social organisation play in determining where environmental limits lie and the rate at which they are approached, together with a warning against the dangers of overestimating the natural component where the social is dominant. Although at times they appear to err in the opposite direction, over-confident of humans' ability to manipulate natural limits, they do acknowledge the possibility of cases where such a course is unavailable. The superiority of their conception lies in its recognition, unlike that of Malthus, of the interrelatedness of the natural, the social and the technological. Thus, whereas Malthus turns the natural element into a rigid and static absolute, Marx and Engels allow a proper consideration of the role of *each* element.

This 'relativised' conception of ecological limits suggests that a comprehensive understanding of the problem requires three avenues of investigation, corresponding to the natural, social and technological elements of ecological limits. To focus exclusively on the natural element would be to take the discoveries of the natural sciences for granted while ignoring their ongoing development. By making natural limits an object of concrete study scientists engage in a social process that in practice relativises those limits, changing nature and its impact by providing us with the knowledge with which to modify our activity. The technological and social elements must also not be studied in isolation, since social factors affect the development and application of technology and technology in turn affects social developments. The interrelation of these elements is at the centre of Marx's theory of historical materialism, the ecological consequences of which will be elaborated in the course of the following chapters.

---

[81] We might even wish to mark such cases by talking of some limits as 'absolute' within a certain context, where that context, for example the context of existing technology, allows no scope for the manipulation of the limit in question. We must remember, however, that the contexts themselves are liable to change over time, rendering the limit in question only a 'relative absolute'. (For a similar terminological device applied to a different subject, see Sayers 1991, pp. 16–17 on the notion of relative truth.)

3 Marxism and the ecological method

It has been a contention of Marx's green critics that Marx makes metaphysical assumptions that are, or uses a method that is, inadequate for comprehending ecological problems. Such an assessment could be invoked on its own, as a free-standing criticism of Marx's method, but is more commonly proposed as an explanation for the alleged ecological insensitivity of Marx's more substantive claims.

The previous chapter demonstrated that simply invoking the notion of 'limits to growth' is not sufficient ground for rejecting Marx's theory. An ecological assessment of historical materialism requires a more subtle analysis of the interactions between nature, society and technology, which together determine the position and character of any limits. The methods of investigation and forms of explanation that Marx brings to bear on this interaction must therefore be subject to scrutiny, alongside the criticisms and counter-claims of his green critics. In order to proceed with this investigation answers to two questions will be required:

(i) To what extent are the methodological and metaphysical claims of green theorists justifiable on philosophical and ecological grounds? and

(ii) How do these claims (insofar as they are justifiable) relate to Marx's theory?

In this chapter I will restrict myself to discussion of the forms of explanation and modes of investigation most appropriate to the study of ecological issues; the more substantive questions about the relation between human societies and non-human nature will be addressed in the following chapter.

3.1 Metaphysical ecology

A useful starting-point for the investigation of green philosophical thought is the survey and criticism given by Andrew Brennan. Brennan's primary objective is to challenge the ecocentric ethic of the Deep Ecology movement while incorporating some of its insights in his own 'ecological humanist' alternative; but in so doing, he makes a study of the metaphysical ideas to which the supporters of Deep Ecology appeal in order to justify their ethical position.

In broad terms Brennan identifies two such ideas. These are: 'idealism (the claim that the world is in some way mind-dependent)', and 'various kinds of global holism (the idea that all things are interdependent in a significant way)'.[1] Brennan interprets these as frameworks of thought transferred from the scientific disciplines of ecology and twentieth-century physics to the more general context of human dealings with nature. Concentrating on frameworks derived from ecology, Brennan draws a useful distinction between 'scientific' and 'metaphysical' ecology. Scientific ecology consists of 'the scientific studies of biologists concerned with the interactions among organisms . . . and between organisms and their environment'; metaphysical ecology, by contrast, is ecology conceived 'as a method of approaching problems, and as supplying a metaphysics that applies to far more than living systems'.[2] Brennan's critique of metaphysical ecology (a label which I will adopt) involves challenges both to the philosophical cogency of its claims and the characterisation of the science from which they are purportedly derived.

It should be noted that though the metaphysical ecologists seek to ground their metaphysic in ecology conceived as a branch of biology, it is not their primary purpose to generalise the results of this science. Rather, their intention is to provide an adequate framework for understanding and dealing with ecological problems and (as we shall see) other problems faced by society. From this perspective the metaphysical ecologists might quite consistently be critical of at least some of the methods and philosophical assumptions of biologists studying natural ecosystems. Even where there is no disagreement, the metaphysical ecologists' use of the scientists' conclusions is more a matter of analogy than generalisation or derivation in a stricter sense. This is acknowledged, for example, by Baird Callicott, who makes it clear that when he speaks of the 'metaphysical implications of ecology' he does not mean to suggest that relations of logical deduction

---

[1] Brennan 1988, p. 7.   [2] *Ibid.*, p. 31.

exist between ecological premises and metaphysical conclusions, and by Arne Naess, who states that metaphysical conclusions 'are not derived from ecology by logic or induction' but rather are suggested or inspired by it.[3] Brennan also argues that the practice of scientific ecology is inconclusive with respect to the claims of metaphysical ecology. Some of these claims, he argues, turn entirely on philosophical considerations, and those which appear to be illustrated, positively or negatively, by scientific ecology require philosophical clarification of the status of ecological investigations. Viewing the relation between scientific and metaphysical ecology as analogical and inspirational rather than logical helps to explain how it is that metaphysical ecology can also draw upon sciences other than ecology. Callicott, for example, draws upon quantum mechanics, as does Fritjof Capra, while Freya Mathews appeals to general relativity in support of her account of the 'ecological self'.[4]

Although Brennan is right to regard idealist views as being common in the writings of metaphysical ecologists, they are by no means universal.[5] Holism, by contrast, may be regarded as the central, defining thesis of metaphysical ecology, as well as being widely cited in more popular and issue-oriented green literature.[6] Typically it is argued that ecological and other problems result from an outdated way of thinking and that a new, holistic scientific paradigm is required in order to understand and solve these problems. This diagnosis is set out in the introduction to Fritjof Capra's influential book *The Turning Point*:

We have high inflation and unemployment, we have an energy crisis, a crisis in health care, pollution and other environmental disasters, a rising wave of crime, and so on. The basic thesis of this book is that these are all different facets of one and the same crisis, and that this crisis is essentially a crisis of perception. Like the crisis in physics in the 1920s, it derives from the fact that we are trying to apply the concepts of an outdated world view – the mechanistic world view of Cartesian-Newtonian science – to a reality that can no longer be understood in terms of these concepts. We live today in a globally interconnected world, in which biological, psychological, social, and environmental phenomena are all interdependent. To describe this world appropriately we need an ecological perspective which the Cartesian world view does not offer.[7]

---

[3] Callicott 1989, p. 101; Naess 1973 quoted in *ibid.*, p. 108.
[4] Capra 1983; Callicott 1989; Mathews 1991.
[5] For example, Callicott (1989, pp. 110–11) takes care to distance himself from such ideas.
[6] See, for example, the references to 'interdependence' in Porritt 1985, p. 3; to 'holism' in Irvine and Ponton 1988, pp. 79, 89; and discussion of 'holistic/visionary greens' in Spretnak and Capra 1986, pp. 3–4 and 140–1. The centrality of such ideas to green politics is commented upon in Dobson 1990, p. 37.   [7] Capra 1983, p. xviii.

Capra is somewhat unusual among metaphysical ecologists in that his primary reason for proposing a new world-view is cognitive; other writers such as Callicott, Mathews and Naess see holistic metaphysics as a way of grounding an environmental ethic. However, in doing so they advance similar claims to Capra, making their arguments equally relevant for the purposes of this chapter.

In order to assess the greens' holism we must try to establish what is intended by the use of this term. Brennan, as noted above, interprets the holism of the metaphysical ecologists as a thesis about the global interdependence of things. This idea is apparent too in the passage quoted above from Capra, and Capra later makes explicit the centrality of ideas of interrelatedness and interdependence to the proposed new paradigm: 'The new vision of reality', he writes, 'is based on awareness of the essential interrelatedness and interdependence of all phenomena – physical, biological, psychological, social and cultural.'[8]

The idea of interdependence is important in green thought of all shades, underlying the greens' advocacy of broad social and economic transformations and hostility to *ad hoc* 'technological fixes' to ecological problems. But this clearly will not suffice, without further elaboration, to ground a new scientific paradigm. Interdependence takes many forms, and indeed Newtonian mechanics, which Capra takes to embody the inadequacies of the old paradigm, was developed precisely in order to explain the causal interconnections between the motions of physical objects.[9] Nevertheless, methodological considerations would appear to be central to the holism espoused by the metaphysical ecologists. The supposed inadequacy of the old paradigm is characterised in writings by Capra, Callicott, Mathews and Skolimowski in terms of the limitations of its 'analytical' and 'reductive' techniques.[10] As Dobson puts it:

---

[8] *Ibid.*, p. 285.

[9] It is sometimes suggested that this 'mechanistic' paradigm is unable to account for the causal feedback loops which play a vital role in the self-regulation of ecological systems. Keekok Lee (1989, p. 53), for example, asserts that 'In classical physics proper, particle C (billiard ball) moves because particle B (another billiard ball) knocks against it, and particle B moves because A (yet another billiard ball) hits it and so on . . . C does not in turn cause A to move.' This, however, is false: such feedback is entirely consistent with classical physics, and patterns of negative feedback are utilised as stabilising mechanisms in many mechanical devices such as the centrifugal regulator which controls the speed of that paradigmatic mechanical system the steam engine. Similarly, James Lovelock, whose investigation of the stabilising effects of negative feedback mechanisms in the biosphere (the Gaia hypothesis) provides much of the context for Lee's discussion, illustrates this phenomenon by reference to a mechanical system: the thermostatic control of an electric oven. See Lovelock 1995, pp. 46–8.

[10] See, for example, Capra 1983, pp. 31–2, 44; Callicott 1989, pp. 102–3; Mathews 1991, pp. 1–3; Skolimowski 1981, p. 27.

The general targets of attack are those forms of thought that 'split things up' and study them in isolation, rather than those that 'leave them as they are' and study their interdependence. The best knowledge is held to be acquired not by the isolated examination of the parts of a system but by examining the way in which the parts interact. This act of synthesis, and the language in which it is expressed, is often handily collected in the term 'holism'.[11]

The holism of the metaphysical ecologists may therefore be interpreted as the view that the degree or type of interconnectedness revealed by ecology (and contemporary physics, etc.) is such as to undermine attempts to understand entities such as ecosystems by means of the reductive techniques of 'Cartesian' science. There is, however, little in the writings cited to indicate how 'reduction' is to be understood, or what reasons there are for regarding it as an inadequate method for investigating ecological problems. These matters will therefore be pursued in the following section.

3.2 Reductionism

'Reductionism', like 'holism', is a term that can refer to a number of different, though related, doctrines. Frequently the charge of reductionism is levelled against a theory as a serious indictment, yet in the history of science the reduction of one theory to another has often been considered an important achievement. I will therefore consider some of the philosophical writings that have sought to clarify what reduction signifies in different contexts and to assess what grounds there may be for asserting or denying its possibility. I will begin with Andrew Brennan's treatment of these questions as they relate to environmental philosophy.

Brennan distinguishes between *terminological* and *ontological* reduction.[12] Terminological reduction (which Brennan holds to be the core philosophical sense of the term) asserts that we can dispense with one set of terms, replacing it with another without any change of meaning. Ontological reduction claims to reduce the number of entities whose existence we acknowledge. Brennan argues that these two theses are independent of each other. On the one hand, he argues, even in a world where just one kind of thing, physical objects, exists, terminological reduction may be impossible; this, he says, is supported by consideration of 'compositional plasticity' in the philosophy of mind: the idea (more commonly referred to as 'variable realisability') that although all mental states have physical instantiations, a particular type of mental state may have different instantiations in different physical systems. On the other hand, Brennan claims,

---

[11] Dobson 1990, p. 37.   [12] Brennan 1988, p. 72.

terminological reduction does not entail ontological reduction, as is shown by the fact that Cartesian dualism – the view that there exist two distinct kinds of entity, the physical and the mental – may be reduced terminologically by substituting 'non-physical' for 'mental'.

From Brennan's examples, however, it is not clear that ontological reduction is, as he claims, independent of terminological questions. The fact that types of mental state (according to the doctrine of variable realisability) are not tied to unique types of physical state, and therefore that names of mental states form an ineliminable part of our descriptions of the world, might reasonably be seen as grounds for asserting that mental states *do* in some sense exist in their own right, notwithstanding their dependence on physical structures. This would be to say that the phenomenon of compositional plasticity shows that types of mental state are *neither* terminologically *nor* ontologically reducible. Brennan could argue against this, by specifying a different conception of what it is for a thing to exist, but in the absence of an alternative account, the supposition that a thing exists if it is ineliminably referred to in our best descriptions of the world seems a reasonable one. More importantly, the lack of a clear account of what it is for a thing to exist, distinct from terminological considerations, suggests that the notion of ontological reduction is unlikely to contribute helpfully to methodological debates.

If 'ontological reduction' is a concept that lacks a clear criterion, then 'terminological reduction' seems to be an empty one from a methodological point of view. Terminological reduction, as defined by Brennan, amounts to no more than the logical equivalence of statements containing different sets of terms. The triviality of his reduction of Cartesian dualism suggests that this form of reductionism is unlikely to yield any substantive methodological precepts. Neither terminological nor ontological reduction, then, seem to address the greens' concerns about the methods of contemporary science. However, Brennan also discusses a third type of reduction which appears to have more in common with the concerns of the metaphysical ecologists.

This we may call *theory* reduction. A paradigmatic example is the reduction of thermodynamics to statistical mechanics. In such cases one theory is deduced from another with the help of additional statements linking the terms of the two theories.[13] For example, the linkage statements needed in order to reduce thermodynamics to statistical mechanics connect

---

[13] These linkage statements are referred to by Nagel (1979, p. 104) as 'bridge laws' and 'rules of correspondence', and by Schaffner (1967, pp. 138, 140) as 'associating sentences' and 'reduction functions'.

statements about the temperature and pressure of a gas with statements about the motions of its molecules.

Brennan's view – albeit contentious – is that reductions of this kind lack any obvious terminological or ontological implications. This conclusion may provide reassurance to any metaphysical ecologists whose opposition to reductive explanation is based on the fear that it will lead to ecosystems and other collective or holistic entities being viewed as less 'real' and there-fore less valuable than the individual organisms that they contain. But even if Brennan's conclusion is correct, it is unlikely to satisfy theorists such as Capra whose concerns are, as we have seen, primarily methodo-logical. The methodological focus of writers like Capra may be seen as an extension of the debate between holists and reductionists (or organicists and mechanists) within biology, a debate which is no longer about the *exis-tence* of a 'life-fluid' or 'life-force' as earlier vitalist theories maintained, but about the most appropriate *method* for studying living organisms.[14] The question for the biologists is whether a science like molecular biology which studies organisms in terms of the 'mechanical' behaviour of their parts can yield the fullest possible understanding of the organism. The metaphysical ecologists address a similar question, but at more than one level: they resist the reductive explanation not only of organisms in terms of their organs or cells, but also of such things as ecosystems in terms of the individual organisms and other entities of which they are composed.[15] Since the metaphysical ecologists' anti-reductionism has this explanatory or methodological character, we may expect analyses of theory reduction to cast light on their claims even if, as Brennan maintains, such reductions lack ontological implications.

The starting-point for many discussions of theory reduction, including Brennan's, is the account given by Ernest Nagel. According to Nagel, one theory is reduced to another when:

(i) the basic terms (and entities) of the reduced theory either appear in the reducing theory or are associated with the basic terms and entities of the reducing theory by means of additional 'associating sentences';

(ii) given the association of basic terms and entities above, the axioms and laws of the reduced theory may be deduced from the reducing theory.[16]

---

[14] See Hull 1974, pp. 127–9.

[15] The fact that Capra regards individual organisms as well as ecosystems as irreducible wholes is illustrated by his citing of holistic medicine as an example of the 'ecological' mode of thought (Capra 1983, p. 118).

[16] Nagel 1961, ch. 11; Nagel 1979, ch. 6. The following account draws upon the survey and criticism of analyses of theory reduction in Schaffner 1967.

Subsequent commentators have proposed variations on this account, according to which, for example, only the observational predictions generated by a theory and not the theoretical entities that it postulates need be accounted for by the reducing theory. Our concern, however, is with the conclusions that Nagel draws from his analysis, and these do not depend upon the disputed points.[17]

Nagel's analysis of reduction was advanced as an account of the way in which science progresses. He was writing within the positivist tradition in which such progress was seen as a process of gradual unification as old theories are absorbed into new more general theories as special cases, and ultimately everything is absorbed into physics. This picture of scientific progress conflicts with the anti-reductionism of the metaphysical ecologists, and Nagel himself rejects metaphysical generalisations of the sort put forward by the metaphysical ecologists, drawn from the alleged irreducibility of one scientific theory to another.[18] His conclusions do, however, allow for a partial accommodation with the anti-reductionism of the metaphysical ecologists.

Nagel holds that since theory reduction is fundamentally a relation between theories rather than between the objects that they describe, and since those theories evolve over time, claims about the reducibility of one theory to another must be specific to particular stages in their development. For example, contemporary thermodynamics is reducible to twentieth-century statistical mechanics but not to that of the eighteenth century; and some, perhaps all, of nineteenth-century chemistry is reducible to post-1925 physics but not to that of a hundred years ago.[19] Nevertheless, it would be mistaken to suppose that we can make no sense of general questions about the reducibility of one domain to another. Such questions can be understood as questions about the future possibility of such reduction, given the scope for development of the two theories. Nagel comes some way towards this position when he writes that unrestricted claims about reduction are most charitably interpreted as recommendations for the direction of future research given the present state of the sciences.[20]

Nagel's view is that such predictions and recommendations must be speculative since we cannot know in advance how the theories will develop and whether a reduction may be achieved in the future.[21] Such a view of the

---

[17] Kemeny and Oppenheim 1956, p. 13. See also Schaffner 1967, p. 138; Hull 1974, pp. 30–1.
[18] See Nagel 1961, p. 337.   [19] *Ibid.*, p. 362.   [20] *Ibid.*, p. 363.
[21] This is also the conclusion reached by Oppenheim and Putnam (1958) on the basis of the Kemeny/Oppenheim definition of reduction. They conclude that unity of science (the idea that a unitary science can be attained through successive reductions) is an empirical thesis for which there is sufficient factual support, in the form of successful reductions at various levels, to justify its acceptance as a working hypothesis.

non-*a priori* character of claims about reducibility is clearly at odds with the unqualified anti-reductionist stance of the metaphysical ecologists. However, there are reasons, some of them acknowledged by Nagel, why in particular cases it may be rational to resist the pursuit of reductionist explanations. It may be, for example, that at a given time the reduction of one science to another would produce few gains in knowledge, in the solution of critical problems in the secondary science, or in guidance for further research, and that energy is therefore better spent pursuing these aims within the secondary science. And even when a reduction has been achieved, the secondary discipline may prove to be the more useful investigative and explanatory tool within its area of application, for example if the lower-level science is too cumbersome for the phenomena in question, requires data that are not available, or fails to suggest fruitful analogies.[22]

It follows from this that the reduction of one science to another need not lead to the elimination of the secondary science, as demonstrated by the continued use of Newtonian rather than Einsteinian physics in the majority of practical contexts. Applying these considerations to ecological issues we reach the conclusion that the pursuit of reductive explanations, even if possible in principle, may not always be the most fruitful approach to dealing with ecological problems. The reducing science, though it may be regarded as more fundamental than the reduced science, and though it may in principle be able to explain the phenomena explained by the reduced science, may in practice be less able to solve the problems of a large and complex system if it can comprehend the processes in this system only by means of lengthy and complex calculations involving extensive initial data whose collection may be difficult if not impossible, or if it focuses attention on the detailed workings of small parts of the system rather than on large-scale processes. The choice between different 'levels' of explanation also has a democratic dimension with relevance to ecological concerns in the need for ecological matters to be debated in terms accessible to a wide audience.

The metaphysical ecologists may thus be right to resist excessive emphasis on explaining ecological and other phenomena in reductive terms, but for the practical and contingent reasons outlined rather than as a matter of immutable metaphysical principle. In the following section I will suggest that Marx and Engels's methodological precepts are broadly consistent with this assessment and that they are not guilty of an excessively reductionist and mechanistic approach, as charged by some of their green critics.

[22] Nagel 1961, pp. 362–3.

3.3 Marxism and method

One element in environmentalist criticism of Marx's method is the claim that his is a mechanistic model of society which ignores or denies the complex causal interconnections of phenomena, as highlighted by the science of ecology. Capra, for example, holds that while Marx had 'profound insights into the interrelatedness of all phenomena' these are contradicted by his technological determinism, and Keekok Lee similarly ascribes to Marx a 'linear model' in which '[i]f the economic base is the cause of the ideological superstructure, then the latter cannot in turn have any effect on the former'.[23]

Such criticisms, however, can be shown to rest upon a misinterpretation of Marx and Engels's writings of a kind which Engels himself explicitly sought to rebut. In a series of letters written during the 1890s Engels dissociates himself and Marx from the idea of a unidirectional causal relation between base and superstructure, insisting that while '[p]olitical, juridical, philosophical, religious, literary, artistic, etc., development is based on economic development', these phenomena 'react upon one another *and also upon the economic basis*'. It is therefore not the case, Engels argues, that 'the economic situation is *cause, solely active*, while everything else is only passive effect'.[24] Moreover, Engels is himself critical of the sort of 'mechanistic' outlook attacked by the metaphysical ecologists, arguing that it is the mechanistic (or 'metaphysical'[25]) assumptions of Marx's critics, and their ignorance of Marx's real method – the dialectical approach derived from Hegel – that lead to the sorts of misinterpretation he is seeking to correct:

What these gentlemen all lack is dialectics. They always see only here cause, there effect. That this is a hollow abstraction, that such metaphysical polar opposites exist in the real world only during crises, while the whole vast process goes on in the form of interaction – though of very unequal forces, the economic movement being by far the strongest, most primordial, most decisive – that here everything is relative and nothing absolute – this they never begin to see. As far as they are concerned Hegel never existed.[26]

---

[23] Capra 1983, pp. 218–19; Lee 1989, pp. 54–5.

[24] 'Engels to W. Borgius in Breslau, London, January 25, 1894', p. 694. First emphasis added. See also 'Engels to F. Mehring in Berlin, London, July 14, 1893', p. 691.

[25] 'Metaphysical' is used here in a sense derived from Hegel, corresponding roughly to the metaphysical ecologists' use of 'mechanistic'.

[26] 'Engels to C. Schmidt in Berlin, London, October 27, 1890', p. 689. See also *Dialectics of Nature*, pp. 173–4, where Engels generalises the view of causal interaction set out in his response to mechanistic interpretations of historical materialism: '*Reciprocal Action* is the first thing that we encounter when we consider matter in motion as a whole from the standpoint of modern natural science . . . Only from this universal reciprocal action do we arrive

There are, of course, passages in Marx and Engels's works – the most prominent being the 1859 Preface – which appear at least consistent with a 'mechanistic' interpretation of historical materialism. Engels responds, however, that such passages merely reflect the priority that he and Marx placed on emphasising the main point at issue between them and their adversaries, namely the primary role of the economic base in explaining history.[27] Such one-sidedness is absent, he argues, from their detailed historical analyses such as Marx's *Eighteenth Brumaire*, 'which deals almost exclusively with the *particular* part played by political struggles and events, of course within their *general* dependence upon economic conditions', and from Marx's discussion in *Capital* of the length of the working day, 'where legislation, which is surely a political act, has such a trenchant effect'.[28] Moreover, recognition of the reciprocal action of the superstructure upon the economic base is integral to the political project that underlies Marx and Engels' work as a whole, since as Engels points out, their struggle for the 'political dictatorship of the proletariat' would be pointless were it the case that 'political power is economically impotent'.[29]

Among contemporary interpretations of Marx, even those that superficially appear closest to the model criticised by Capra and Lee reject the idea of a unidirectional causal link between base and superstructure. G. A. Cohen's analytical reformulation of Marxism rejects the language of dialectics, used by Engels in the passages cited above, and asserts the 'explanatory primacy' of the development of the productive forces within historical materialism, and, for this reason, has been criticised by other Marxists as 'mechanistic',[30] yet his conception of the functional relationship between base and superstructure not only allows for, but requires the superstructure to affect the base. Superstructures, according to Cohen, are 'selected' according to whether they *sustain* or *frustrate* the economic forms that constitute the base, and these forms themselves rise and fall according to whether they promote or inhibit the development of the productive forces.[31] Since Cohen's concept of selection builds upon Marx's own account (to be discussed in chapter 5 below) of the way in which the economic forms and political superstructures brought into existence by the development of the productive forces are overthrown when they become

footnote 26 (*cont.*)
at the real causal relation. In order to understand the separate phenomena, we have to tear them out of the general inter-connection and consider them in isolation, and there the changing motions appear, one as cause and the other as effect.'
[27] 'Engels to J. Bloch in Königsberg, London, September 21[-22], 1890', p. 683.
[28] 'Engels to C. Schmidt in Berlin, London, October 27, 1890', p. 689.
[29] *Ibid.*   [30] E.g. Sayers 1984, pp. 10–11. Cf. Carling 1991, p. 16.   [31] E.g. Cohen 1988, p. 10.

'fetters' on their further development, it is hard to see what grounds there could be for ascribing to Marx a 'linear' or unidirectional model of causation.

Turning from Marx and Engels's substantive model of society to their explicitly methodological pronouncements, we can see certain parallels with the claims of metaphysical ecology. In some of his later writings Engels sought to place his and Marx's account of society within a broader philosophical and methodological context, and while much of his discussion suffers from the same vagueness as that of the metaphysical ecologists (exacerbated in some cases by the language of dialectics), he may also be seen to anticipate many of their insights, and on occasion with greater sophistication and clarity.

The first point of comparison is Engels's conception of the dialectical method as being concerned with the interconnections of the objects of study: he defines 'dialectics' as 'the science of interconnections',[32] and argues that, in contrast with 'metaphysics', it 'grasps things, and their images, ideas, essentially in their interconnection, in their sequence, their movement, their birth and death'.[33] Like the metaphysical ecologists, Engels contrasts this approach with approaches that overemphasise the division of nature into its component parts:

The analysis of Nature into its individual parts, the grouping of the different natural processes and natural objects in definite classes, the study of the internal anatomy of organic bodies in their manifold forms – these were the fundamental conditions of the gigantic strides in our knowledge of Nature which have been made during the last four hundred years. But this method of investigation has also left us as a legacy the habit of observing natural objects and natural processes in their isolation, detached from the whole vast interconnection of things; and therefore not in their motion, but in their repose; not as essentially changing, but as fixed constants; not in their life, but in their death. And when, as was the case with Bacon and Locke, this way of looking at things was transferred from natural science to philosophy, it produced the specific narrow-mindedness of the last centuries, the metaphysical mode of thought.[34]

Engels's view of this analytical approach as performing a necessary function in the development of science, but harmful if emphasised to the exclusion of more integrating modes of thought, is strikingly similar to Capra's characterisation of such techniques as useful and in some cases necessary, but dangerous when taken to constitute a complete explanation.[35] Capra

---

[32] *Dialectics of Nature*, p. 26.   [33] *Anti-Dühring*, p. 27.
[34] *Ibid*. See also *ibid.*, p. 28, and *Ludwig Feuerbach and the End of Classical German Philosophy*, pp. 55–6.   [35] Capra 1983, p. 288.

also follows Engels in viewing the extension of such techniques beyond the natural sciences as especially problematic.[36]

Engels also offers an account of scientific reduction which recognises the possibility that lawlike connections will be established between the phenomena (or 'forms of motion') explained by 'higher' and 'lower' sciences, but without assuming that this will enable us to dispense altogether with the higher sciences. He is critical of the natural scientists' 'craze to reduce everything to mechanical motion . . . which obliterates the specific character of the other forms of motion', but continues:

This is not to say that each of the higher forms of motion is not always necessarily connected with real mechanical (external or molecular) motion, just as the higher forms of motion simultaneously also produce other forms; chemical action is not possible without change of temperature and electric changes, organic life without mechanical, molecular, chemical, thermal, electric, changes, etc. But the presence of these subsidiary forms does not exhaust the essence of the main form in each case. One day we shall certainly 'reduce' thought experimentally to molecular and chemical motions in the brain; but does that exhaust the essence of thought?[37]

This account, schematic as it is, shares the metaphysical ecologists' antipathy to an overstated reductionism, but appears consistent with the conclusions drawn from Nagel's analysis of reduction, discussed above. Relatedly, we can see in Engels's illustration of the dialectical 'law' of the transformation of quantity into quality, an endorsement of the doctrine of emergent properties:

If we imagine any non-living body [Engels states that the same holds for living bodies, but under more complex conditions] cut up into smaller and smaller portions, at first no qualitative change occurs. But this has a limit: if we succeed, as by evaporation, in obtaining the separate molecules in the free state, then it is true that we can usually divide these still further, yet only with a complete change of quality. The molecule is decomposed into its separate atoms which have quite different properties from those of the molecule . . .
But the molecule is also qualitatively different from the mass of the body to which it belongs . . .
Thus we see that the purely quantitative operation of division has a limit at which it becomes transformed into a qualitative difference: the mass consists solely of molecules, but it is something essentially different from the molecule, just as the latter is different from the atom.[38]

In endorsing the idea that composite bodies can possess properties distinct and unpredictable from the properties of their parts, Engels is at one with the metaphysical ecologists, but he recognises more clearly than they do that there is nothing mysterious or unique to the biological sciences in

[36] Ibid., p. 55   [37] Dialectics of Nature, pp. 174–5.   [38] Ibid., pp. 28–9.

emergent properties, and that their existence at any level of organisation is a matter of empirical determination.[39]

We have seen that there is much in Engels's methodological pronouncements that anticipates what is valid in the views of metaphysical ecologists such as Capra. Capra himself, as noted above, holds that there is a tension in Marx between an 'organic' or 'systems' view, corresponding in a general way to the holistic outlook of the metaphysical ecologists, and another, more deterministic tendency. The latter, Capra suggests, was intended by Marx to enhance his theory's scientific credibility, but tends, along with his habit of 'expressing his ideas in "scientific" mathematical formulas' to undermine the larger 'organic' picture.[40]

Why Marx's use of mathematical formulas should be regarded as problematical is unclear. Presumably Capra would characterise it as an application of the techniques of Cartesian science beyond their legitimate field, yet numerical methods are used in both scientific ecology and quantum mechanics from which Capra derives his ecological paradigm, and there is nothing in Capra's work to indicate why such methods should not be applied to the study of society. Capra's claim that there is a contradiction between Marx's technological determinism and his holism is also misconceived. He finds problematic not only the sort of mechanistic picture from which Engels dissociates himself and Marx, but also any suggestion that technology is a predominant influence on society, counterposing this to an organic perspective in which ideology and technology are equally important. Such a perspective, however, is clearly contrary to Engels's presentation of historical materialism as discussed above, and is also alien to Marx. An examination of the methodological principles outlined by Marx and his application of these principles in his political economy will show that his approach does have a holistic character broadly resembling that of metaphysical ecology, but that the kind of holism espoused by Marx does not exclude there being an element which is predominant in determining the behaviour of the system as a whole.

In the *Grundrisse*, Marx counterposes two approaches to the study of economics:

The economists of the seventeenth century, e.g., always begin with the living whole, with population, nation, state, several states, etc.; but they always conclude by discovering through analysis a small number of determinant, abstract, general relations such as division of labour, money, value, etc. As soon as these individual

---

[39] Cf. Nagel 1961, pp. 366–74; Brennan 1988, pp. 86–8.
[40] Capra 1983, pp. 216–17, 218–19.

moments had been more or less firmly established and abstracted, there began the economic systems, which ascended from the simple relations, such as labour, division of labour, need, exchange value, to the level of the state, exchange between nations and the world market. The latter is obviously the correct scientific method.[41]

Marx thus indicates that the highest task of economics is not the analytical breakdown of the whole into simple parts, but the synthesis of the parts in order to reconstruct the whole, revealing the relations that make up its internal structure. Marx does not, however, deny the need for the analytical process; in the passage above it is clear that he regards the identification and firm establishment of the 'individual moments' as a precondition for the synthetic phase.[42] In the *Critique of Political Economy*, and *Capital*, Marx begins by examining one element of the capitalist economy, the commodity; he examines its internal structure and its relations with other elements of the economy in order to build up a picture of the totality. But the selection of the commodity as a starting-point is neither arbitrary nor obvious, and it is only at the end of Marx's preliminary work, the *Grundrisse*, after the chapters on money and capital, that, in a section marked 'to be brought forward', Marx identifies the commodity as 'the first category in which bourgeois wealth presents itself'.[43]

Marx's method in *Capital* has been called the ascent from the abstract to the concrete,[44] but this definition is only accurate if we understand 'abstract' in a particular way. The commodity is not an abstraction in the empiricist sense of a concept arrived at by separating out what is common in a multiplicity of different objects. For Marx the commodity is a part of the material world, which 'can only exist as an abstract, one-sided relation within an already given, concrete, living whole'.[45] It is, in other words, a part which is internally related to the whole. The whole itself, for Marx, is concrete in that 'it is the concentration of many determinations, hence unity of the diverse',[46] a system made up of many interrelated parts.

The scientific process described by Marx therefore consists of rebuilding the concrete in thought, out of the parts into which the concrete in reality

---

[41] *Grundrisse*, p. 100.

[42] Elster (1985, pp. 121–2) makes a similar point regarding Marx's use of economic 'models' in which certain aspects of the economic system are temporarily ignored or taken out of context in order to facilitate understanding. This 'method of successive approximations', Elster writes, enables Marx to avoid 'the trap of premature totalization which from Lukacs onward has plagued Western Marxism'.

[43] *Grundrisse*, p. 881. As we will see, 'first' in this context does not indicate historical priority, but rather that the commodity is the form of capitalist wealth most important in shaping the economic system.   [44] Ilyenkov 1982. See also Bakhurst 1991.

[45] *Grundrisse*, p. 101; cf. Ilyenkov 1982, p. 76.   [46] *Grundrisse*, p. 101.

has been decomposed. But why should it matter to Marx which 'abstract determination' is taken as its starting-point? If all the economic categories are interconnected within the concrete whole, could we not start with any category and by exploring its relations reach an understanding of the whole? Such an answer would correspond to Capra's notion of an organic whole as a system in which no element is dominant, but it is firmly rejected by Marx. The order of the categories, he argues, must be determined by their relations within the economic structure, in which some predominate over others. For this reason Marx argues against a historical approach to the ordering of the categories, rejecting (as Hegelian) the suggestion that the sequence in which the different economic forms emerge is the same as that dictated by their relations in mature capitalist society:

For example, nothing seems more natural than to begin with ground rent, with landed property, since this is bound up with the earth, the source of all production and of all being, and with the first form of production of all more or less settled societies – agriculture. But nothing would be more erroneous. In all forms of society there is one specific kind of production which predominates over the rest, whose relations thus assign rank and influence to the others. It is a general illumination which bathes all the other colours and modifies their particularity. It is a particular ether which determines the specific gravity of every being which has materialized within it.[47]

In bourgeois society, Marx continues, capital is the dominant form of production, and it turns agriculture more and more into a mere branch of industry. Hence capital must logically precede ground rent:

Ground rent cannot be understood without capital. But capital can certainly be understood without ground rent. Capital is the all-dominating economic power of bourgeois society. It must form the starting-point as well as the finishing-point, and must be dealt with before landed property. After both have been examined in particular, their interrelation must be examined.[48]

In the Introduction to the *Grundrisse* Marx takes the argument no further, but later, as we have seen, a similar form of reasoning is applied to determine that the investigation of capital itself must begin with the analysis of the commodity. The commodity is the 'elementary form' of capital, the form in which the wealth of capitalist societies appears.[49] It is the commodity form, the unity of value and use-value, which makes possible surplus value, and this in turn makes possible profit. Profit cannot be understood without the commodity, but the commodity can be understood, in itself and in its simple exchange relations, without profit.

I have suggested that the methodological principles advanced and

---

[47] *Ibid.*, p. 107.   [48] *Ibid.*   [49] *Capital*, vol. I, p. 125.

applied by Marx and Engels may be seen to incorporate the main insights of the metaphysical ecologists (recognition of the interrelatedness and interdependence of phenomena and the dangers of over-zealous reduction), while being in some respects more developed. General questions about the extent to which Marx's and Engels's methodological views coincide are too large to be considered here, but, to summarise what has already been said, we may note that both of them regard the object of investigation as a complex 'organic' whole, comprising many internally related parts, into which it cannot be mentally broken down without doing violence to the integrity of the whole and to those of its parts that cannot exist in reality apart from their relation to the whole. Such analysis is, for Marx and Engels, a necessary condition for overcoming the opacity of the whole, but is not sufficient since the whole must be rebuilt transparently in thought, by understanding the relations between the parts. For both of them this leads to a view of the object as a complex system whose nature and development is determined not by a single cause but by the interaction of various elements; an interaction in which, however, one element may predominate if it is critical in determining the conditions under which the interaction takes place.

This argument would be incomplete, however, if we did not consider the case for a methodological individualist interpretation of Marx, put forward by Jon Elster. Superficially, Elster's arguments may seem distant from the concerns of the metaphysical ecologists, since the focus of his argument is not holism as such, but functional explanation, about which the metaphysical ecologists have little to say. However, as Elster makes clear, there is an association between functional explanation, at least in the forms that he finds most objectionable, and the rejection of methodological individualism.[50]

To give a functional explanation of something is to explain its existence by pointing out that it has a certain function, where 'function' is understood as a consequence that is beneficial to some entity. To take a standard example, we may explain the fact that a bird has hollow bones by pointing out that hollow bones facilitate flight, which benefits the bird by increasing its chances of survival and reproduction. Such explanations reverse the usual pattern of causal explanations in which effects are explained by their causes. However, we find the functional explanation acceptable in this case because we have an account – namely Darwin's theory of natural selection – which explains in causal terms how it is that species tend to acquire char-

---

[50] Elster 1985, pp. 6–7.

acteristics that confer reproductive and survival advantages. Marx, according to G. A. Cohen's view (to be considered further in chapter 5), gives a functional explanation of the existence of particular relations of production at a given time and the transition from one set of relations to another: the relations prevailing at any time are to be explained by their tendency to promote the development of the productive forces. Elster holds, however, that functional explanations of this kind are not legitimate in the social sciences.

Elster gives differing accounts of why functional explanation is to be rejected. In 'Marxism, Functionalism, and Game Theory' he writes that functional analysis 'has no place in the social sciences, because there is no sociological analogy to the theory of natural selection'.[51] The latter claim, however, has been challenged,[52] and in *Making Sense of Marx* Elster's argument appears to be not that it is impossible to specify a causal mechanism underpinning functional explanation in the social sciences, but that Marx did not provide such mechanisms to support *his* use of functional explanation, and that where such a mechanism *is* supplied (as it must be to make the functional explanation valid) the functional explanation is rendered redundant.[53]

On either account, the principle underlying Elster's critique of functional explanation in the social sciences is that functional explanations are only valid when conjoined with an account of the causal mechanism leading from the beneficial consequence back to the existence of the phenomenon to be explained His methodological individualism consists in the further requirement that the mechanism be at the level of individual human beings, methodological individualism being defined as 'the doctrine that all social phenomena – their structure and their change – are in principle explicable in ways that only involve individuals – their properties, their goals, their beliefs and their actions'.[54] This, as Elster notes, is a reductionist perspective, the explanation of social phenomena in terms of individual behaviour being analogous to the explanation of cells in terms of molecules.[55] Elster's perspective therefore appears fundamentally opposed to that of metaphysical ecology. Moreover, moving from analysis to exposition, Elster locates Marx's methodological contribution to the social sciences not in his use of functional explanation (which according to

---

[51] Elster 1989, p. 65 (first published in 1982).
[52] See, for example, Van Parijs 1982, pp. 498–501. The causal mechanism implicit in Cohen's account, and alternatives to it, will be discussed in chapter 5.
[53] Elster 1985, pp. 28–9. This is also the interpretation of Elster's argument adopted by Alan Carling (1991, pp. 17–24) in his useful discussion of the Elster–Cohen controversy.
[54] Elster 1985, p. 5; 1989, p. 48.   [55] Elster 1985, p. 5.

Elster is to be deprecated) but in an account of the way in which history is shaped by the unintended consequences of individual human actions, an account which in Elster's view is consistent with methodological individualism and can be elaborated using the tools of rational choice theory and game theory.[56] It would seem, therefore, that Elster's theory allows little room for the sort of reconciliation between Marx and metaphysical ecology sketched above. However, there are reasons relating to Elster's account for resisting this conclusion.

First, we may reject the view that functional explanations always require specification of the underlying causal mechanism in order to be valid. Elster is right to insist that evidence is necessary to distinguish genuine cases of functional explanation from the many cases in which the beneficial consequences of some phenomenon play no part in explaining its existence. This, however, need not involve detailed specification of the causal mechanism, since we may have empirical reasons for thinking that such a mechanism must exist without knowing what it is. As Cohen argues, 'one can support the claim that B functionally explains A even when one cannot suggest what that mechanism is, if instead one can point to an appropriately varied range of instances in which whenever A would be functional for B, A appears'.[57] It may therefore be possible to explain composite entities and their behaviour functionally, without specifying the underlying causal mechanism at the individual or any other level. This possibility, however, offers little in the way of concessions to the metaphysical ecologists since it remains the case that functional explanations are rendered more complete and more secure by specification of the underlying causal mechanism.

The metaphysical ecologists may take more comfort from a critique of the arguments underpinning Elster's reductionism. Elster gives two basic reasons for seeking explanatory mechanisms at the individual level. Firstly, he claims that causal relationships postulated at this level are less prone to error than those postulated at higher levels of aggregation, with less chance of confusing genuinely explanatory relationships with chance correlations. Secondly, he suggests that the provision of mechanisms at the individual level is a *part of* what it is to explain something: '[t]o explain is to provide a *mechanism*, to open up the black box and show the nuts and bolts, the cogs and wheels, the desires and beliefs that generate the aggregate outcomes'.[58] Note, however, that these considerations point only to a

---

[56] Elster 1985, ch. 1, especially pp. 3–4; Elster 1989, especially pp. 66–7.
[57] Cohen 1989, p. 98. See also Cohen 1978, ch. 9; cf. Elster 1985, pp. 28–9.
[58] Elster 1985, p. 5.

difference of degree between explanations posited at the level of the individual and at higher levels of aggregation, and not to any absolute privileging of the former. Moving down the reductionist scale may *reduce* the scope for confusing causation with mere correlation[59] but does not eliminate it; and the actions and choices of individuals, while explanatory in relation to the behaviour of classes and other collective entities, may themselves be considered 'black boxes' in relation to explanations that might in principle be offered by neuroscience.[60]

The implausibility of finding useful explanations of social processes or even individual actions at the neurological level may remind us of the point made by Nagel, that reduction to the lowest possible level may not always be the most productive avenue of research. Elster acknowledges that 'black-box' explanations may be preferable to 'premature reductionism' in cases where the latter is likely to yield 'sterile and arbitrary explanations', but insists that this methodological collectivism 'can never be a desideratum, only a temporary necessity'. We need not accept, however, that the provision of micro-foundations always renders the higher-level explanation redundant. An alternative view, closer to that of Nagel, is suggested by John Roemer who shares Elster's view that we need to find individual-level micro-foundations for social phenomena such as class but writes that '[t]he most efficacious lens for analysis may not be the one with the highest magnifying power, and by resolving always to the level of the individual one may lose the pattern'.[61] For Roemer, explanations at the level of the individual are necessary not to supersede class analysis but 'to explain why classes are the relevant unit of analysis'.[62]

3.4 Conclusion

In this chapter I have adopted a deflationary approach to the claims of the metaphysical ecologists, arguing that their claims, insofar as they hold true, amount in the main to fairly straightforward assertions of causal interconnectedness which are in no way ruled out by the principles of classical physics or mechanics. And while the complex interconnectedness of natural and social systems may frustrate hopes of explaining them reductively, in terms of their constituent 'atoms', this does not licence the rejection of such explanation as a matter of principle encountered in much of the metaphysical ecology literature. It is clear, nevertheless, that scientists

---

[59] But for an opposing view, see Roberts 1996, p. 22.
[60] For a related argument, see Carling 1991, p. 22.
[61] Roemer 1982, p. 519; cf. Nagel, text to note 22 above. [62] Roemer 1982, p. 513.

and others have often failed to appreciate fully the degree to which parts of the biosphere are interdependent, and to this extent the observations of the metaphysical ecologists may serve a useful heuristic function.

I have argued that Marx and Engels recognise the forms of interdependence and interconnectedness to which the metaphysical ecologists draw attention, both in their explanations of social phenomena (historical materialism) and in their methodological reflections and practice. The extent to which this is carried through into their account of the relation between humanity and nature will be considered in the next chapter, but to anticipate a little, Engels's examples of ecological problems in the following passage usefully illustrate the dangers of underestimating the extent to which different elements of the ecosystems upon which humans depend are causally interdependent.

The people who, in Mesopotamia, Greece, Asia Minor and elsewhere, destroyed the forests to obtain cultivable land, never dreamed that by removing along with the forests the collecting centres and reservoirs of moisture they were laying the basis for the present forlorn state of those countries. When the Italians of the Alps used up the pine forests on the southern slopes, so carefully cherished on the northern slopes, they had no inkling that by doing so they were cutting at the roots of the dairy industry in their region; they had still less inkling that they were thereby depriving their mountain springs of water for the greater part of the year, and making it possible for them to pour still more furious torrents on the plains during the rainy seasons. Those who spread the potato in Europe were not aware that with these farinaceous tubers they were at the same time spreading scrofula.[63]

Engels's examples are outdated in one respect: they do not reflect the global scale of environmental interdependence illustrated by many contemporary environmental problems, such as radioactive discharges which can contaminate food supplies and cause cancers on distant continents, and emissions of gases that can cause global climate change or damage the ozone layer on the other side of the world. One aspect of the interdependence of ecological systems that is illustrated by all of these examples, however, is the way in which ecological problems cut across traditional divisions between the sciences. Engels's examples show the need for knowledge of agriculture and forestry to be combined with knowledge of hydrology and medicine. The more recent examples combine physics and chemistry with meteorology, climatology, ecology, etc. The extent to which ecological problems cross scientific divisions is illustrated by the interdisciplinary nature of much environmental research and the range of disciplines contributing to it, including all those mentioned above as well as

---

[63] 'The Part Played by Nature in the Transition from Ape to Man', p. 362.

biology, geography, economics, sociology, political science, and many others. What this points towards is not a unification of science in the positivist sense, in which everything is reduced to physics, but a recognition of the continuity and overlap of the subject matter of the various disciplines and the need to allow knowledge acquired in one discipline to be fed back into another.

The most pronounced division in traditional science to be bridged by ecological problems, and the most important for the purpose of designing policies to combat them, is the division between the natural and the social sciences. The existence of serious ecological problems with the potential to undermine the sustainability of human societies undermines any view of the social sciences as independent of the natural sciences; and conversely, the natural sciences are forced to take account of the impact of humans upon their objects of study, and consequently of the social sciences that study the determinants of that impact.[64] What is needed, therefore, is a theory of society which recognises the role that natural conditions play in shaping society, and the ways in which society itself can affect those conditions. Marx's historical materialism is a theory which on some interpretations claims to do this, and in the following chapter I will consider whether it succeeds.

---

[64] Recall that Odum, quoted at the beginning of chapter 1, reports the increasingly interdisciplinary nature of ecology, including its incorporation of social sciences. Consider also the natural sciences referred to above: hydrology, climatology, meteorology.

4 Historical materialism: locating society in nature

In chapter 2, I rejected the Malthusian notion of nature as an absolute constraint on social development, on the grounds that nature's impact is always mediated by social and technological factors. However, the fact that nature has an impact, and that the scope for mediation of that impact is limited, cannot be denied. We have seen that Marx and Engels recognise these facts in various of their statements, including their criticisms of Malthus. It may be argued, however, that this is mere lip-service; what is needed is not simply that these facts be recognised, but that they be properly theorised and incorporated into a political programme. My view is that Marxism has the resources for such a theorisation, and in this chapter I take a further step in arguing for this view, by showing how the idea of human dependence upon nature forms an integral and central component of Marx's theory of society.

4.1 Ecology and human dependence upon nature

Let me begin by stating the obvious: the relation between human beings and non-human nature is a two-way affair. Humans are affected by non-human nature and in turn affect it. Indeed, the two elements of this relation, and their interplay, are essential to our understanding of environmental problems. If humans were not affected by nature, then the disruption of natural systems, whether by human action or by natural processes, would not produce environmental *problems*, at least as they are usually understood and as they were defined in chapter 1. The fact that we do regard certain environmental changes as problems is indicative of the effects that they have upon us. These may range from the narrowly instrumental – for example, the effect of atmospheric ozone depletion upon human health and agricultural production – to the more rarefied effects that destruction of wilderness or habitats may have upon our aesthetic or moral sensibilities, but the phenomena most widely perceived as environmental problems are those which threaten to have the most profound negative consequences for human beings. If, on the other hand, humans had no impact on nature, then we would have no cause to fear human-induced environmental problems, but by the same token could do nothing to avert natural disasters or to improve our environment. Clearly both these suggestions are absurdities, and a viable account must recognise that this is a reciprocal relation, in which each side, humanity and nature, is both cause and effect.

Given the holistic leanings of many green theoreticians, and particularly their endorsement of the form of holism that consists in the recognition of complex and reciprocal patterns of causation, we would expect a recognition of this reciprocity in the human–nature relation. Indeed, green criticisms of contemporary society do address both aspects of this relation: the harmful effects of humans upon nature are catalogued and condemned, and their possible repercussions are highlighted as a deadly warning of where our actions may lead us. The latter aim, however, and the concomitant emphasis on the vulnerability of human beings, sometimes leads to a one-sided presentation of human–nature interaction. Contemporary society is rightly criticised for giving inadequate attention to its dependence upon nature, but affirmation of this dependence is sometimes equated with a denial of human influence over nature. Thus it is assumed that thinkers who, like Marx, strive for greater human control over nature, must underplay human dependence upon nature. This, however, ignores the fact (underlying Grundmann's defence of the domination of nature, discussed in chapter 1) that efforts to increase human control over nature may not only be conducted with an awareness of their possible ecological repercussions arising from human dependence upon nature, but may be motivated precisely by a desire to reduce the negative repercussions of that dependence, such as the scarcity of means to satisfy their needs or their vulnerability to natural disasters.[1]

The idea that human dependence on nature implies a lack of influence over nature is also apparent in the many green writings that, with some justification, question the capacity of science and technology to solve ecological problems. This downplaying of humans' ability to manipulate

---

[1] This observation constitutes what is correct in Grundmann's defence of the domination of nature; his mistakes, as outlined in chapter 1, lie in his definition of 'domination' and his supposition that an instrumentalist and interventionist relation to nature, which the metaphor of domination tends to imply, can account for all that is of value in it.

nature may appear to contradict that other tenet of holistic environmental philosophy, the insistence that human actions have far-reaching effects 'throughout the system'. The paradox, however, is only apparent. Its solution lies in the distinction between conscious modification of nature in accordance with human interests, and those effects of human action which are unforeseen or at any rate not desired. Unforeseen or undesired consequences of human action, where they prove detrimental to human interests, can constitute environmental problems. An ecologically aware account, therefore, should not downplay the ability of humans to influence their environment, but must recognise the gap between the intentions that lie behind environmental interventions and the effects of those interventions.[2]

The purpose of the foregoing discussion has been to illustrate how recognition of the reciprocity of the human–nature relation is a necessary condition for a theoretical framework within which ecological problems can be understood. In order to apply this understanding to the examination of Marx's work (or indeed any other social theory), it will be useful to label the two sides of this relation in the following way.

(i) The principle of *ecological dependence* states that humans are dependent upon nature for their survival, and consequently for anything else that they may wish to do, and that the characteristics of the nature they confront have significant causal impact upon the course of their lives.

(ii) The principle of *ecological impact* states that human actions have significant effects (planned and unplanned) upon nature.

It may be thought that in separating the two sides of the human–nature relation in this way I am adopting the sort of 'mechanistic' approach whose criticism by the greens was, with some reservations, endorsed in the last chapter. This, however, is not the case. My purpose in stating these principles is to facilitate examination of the way in which the human–nature relation is understood by Marx and by his critics. I do not mean to suggest that the two principles can be realised independently of each other; indeed I stated above that such a suggestion is absurd and sought to show that, contrary to what is sometimes assumed, the principles can and must co-exist in a satisfactory theory.

Another aspect of the human–nature relation that is widely held to be

---

[2] Unintended consequences can be a result of lack of knowledge (unforeseen consequences), or they can be consequences that are foreseen but are not the intended purpose of the action. In the latter case, net harm may be done either because of a conflict of interests (externalities) or because the structure within which the decision is made excludes certain (e.g. long-term or non-monetary) considerations from the calculus.

important may be expressed by adding another principle to the two already given:

(iii) The principle of *ecological inclusion* states that humankind is a part of nature.

I state this third principle with some reservation, since it is not clear without further specification what this statement adds to the previous two. The statement that humankind is a part of nature may refer to several aspects of the human–nature relation which do not necessarily coincide. It may mean that the human species emerged from a pre-existing nature, which is true but does not in itself tell us anything about the contemporary relation between humankind and nature or the implications of this relation. It may be a statement of a philosophical monism, according to which *everything* is a part of nature. This again has little impact on issues of ecology, at least until the operative concept of nature has been fleshed out. Or more usefully, it may assert of a degree of similarity, or of causal interaction between human beings and other entities. On this interpretation, however, it is not clear how far the statement that humans are *included* in nature differs from the claim that, like other species, they are *dependent*, and have an *impact,* upon (other parts of) nature. As I suggested at the end of chapter 3, the fact that society and nature interact would seem to suggest some continuity of the objects and of the science that describes them. Because of this ambiguity in the third principle I will concentrate my attention on the first two. However, since the third principle is widely affirmed in environmental writings it will be worthwhile to record the extent to which it is reflected in Marx's works.

It may have been noticed that the complementary principles of ecological dependence and ecological impact are not exactly analogous to one another. So far as causal influence is concerned each side affects the other, but the *existential* dependence of humans upon nature is a unilateral affair. Matter, both as external nature and as the human body that has evolved out of it, must exist, and exist in one of a limited range of configurations, as a necessary condition for the existence of living, thinking human beings. Conversely, nature existed before there were humans, and will certainly exist in their absence in the future. The principle of ecological dependence may therefore be subdivided into (a) the principle of *causal dependence*, and (b) the principle of *existential dependence*.

As I have indicated, Marx has been accused of neglecting the principle of ecological dependence. In the following sections I shall examine and criticise some of the arguments that have given rise to this charge.

However, since there is a popular perception – driven by revelations of the environmental damage that has been wrought in countries whose governments professed to be guided by Marxism, and reinforced by the perception of Marxism as a form of economic or technological determinism – that Marxism can have nothing, or at least nothing helpful, to say about the role that nature plays in the life of human societies, I will begin by presenting the *prima facie* case in Marx's favour. The case, in essence, is that Marxism *can* be an appropriate framework for the investigation of ecological problems in virtue of the central part played by the notion of human dependence upon nature in Marx's materialist conception of history.

4.2 Marx's materialism and human dependence upon nature

Marx describes his theory of history as materialist, but the term 'materialism' can mean various different things. In its strongest sense materialism is the claim that all that exists is matter. Marx, however, did not hold this view, for as Keith Graham points out, his analysis of the commodity is explicit in attributing to it a non-material or 'supra-natural' property, namely value.[3] A weaker sense of 'materialism', more plausibly attributed to Marx, is the claim that everything that exists is, *or is dependent upon*, matter. But while this form of materialism has a certain affinity with the principle of ecological dependence, particularly that aspect of it which asserts a unilateral existential dependence of humanity upon nature, it is too abstract to have any direct ecological implications. It says nothing about what matter it is that humans depend upon, and is therefore consistent with a world in which each person's existence depends upon the existence of his or her body but not upon any other material object. It also says nothing about the causal impact that natural phenomena can have upon humanity even when the conditions for human existence are satisfied. This does not mean, however, that Marx's materialism is irrelevant to the principle of ecological dependence, for Marx's description of his theory of history as materialist adverts not primarily to the philosophical doctrine just described, but to a cluster of 'materialist' explanatory claims that he advances, from within, but not essentially related to, this general philosophical perspective.[4] It is in this cluster of explanatory claims that Marx's commitment to the principle of ecological dependence is to be found.

One writer who anticipates the account of Marx's materialism to be presented here is Sebastiano Timpanaro. Arguing in 1966 against contempo-

---

[3] Graham 1992, p. 9; cf. *Capital*, vol. I, pp. 128, 149.
[4] Cf. Marx's own reference to his materialist method in *Capital*, vol. I, p. 494n.

rary Marxists who had dismissed nature as a social construction, he writes that Marx's materialism is to be understood 'above all' as an

acknowledgement of the priority of nature over 'mind', or, if you like, of the phys-ical over the biological level, and of the biological level over the socio-economic and cultural level; both in the sense of chronological priority (the very long time which supervened before life appeared on earth, and between the origin of life and the origin of man), and in the sense of the conditioning which nature *still* exercises on man and will continue to exercise at least for the foreseeable future.[5]

In a graphic metaphor of his own, Timpanaro ridicules those who endorse Marx's base–superstructure metaphor as an expression of the dependence of society's legality, polity and ideology upon its economic forms, but who reject the dependence of these forms upon natural preconditions. 'The position of the contemporary Marxist', he writes,

seems at times like that of a person living on the first floor of a house, who turns to the tenant of the second floor and says: 'You think you're independent, that you support yourself by yourself? You're wrong! Your apartment stands only because it is supported on mine, and if mine collapses, yours will too'; and on the other hand to the ground-floor tenant: 'What are you saying? That you support and con-dition me? What a wretched illusion! The ground floor exists only in so far as it is the ground floor to the first floor. Or rather, strictly speaking, the real ground floor is the first floor, and your apartment is only a sort of cellar, to which no real exis-tence can be assigned.'[6]

Although these passages may be seen implicitly to endorse the principle of ecological dependence and to recognise the asymmetry of humans' exis-tential dependence upon non-human nature, Timpanaro's own concerns are not specifically ecological. He is concerned, firstly, with the danger that the 'idealist' tendencies he discerns in his contemporaries may provoke a 'vulgar-materialist' and possibly racist backlash,[7] and, secondly, to under-stand the biological determinants of the human condition in general (human nature) and of individual human lives. A key issue for Timpanaro is the problem of 'physical ill', and the fact that it cannot be ascribed solely to bad social arrangements but has a 'zone of autonomous and invincible reality'.[8] A parallel may be drawn between the view that Timpanaro is attacking here and the attempts of some – including some Marxists – to blame ecological problems solely upon social conditions, ignoring the material factors that will tend to give rise to them under any social arrange-ments. More generally, Timpanaro's reassertion of the existence of a 'zone of autonomous and invincible reality' remains necessary today in the face of loose assertions by post-modernists of the 'social construction of

[5] Timpanaro 1975, p. 34.   [6] *Ibid.*, p. 44.   [7] *Ibid.*, pp. 13, 17.   [8] *Ibid.*, p. 20.

reality'.[9] Timpanaro thus points us in the direction of an 'ecological' interpretation of Marx's materialism, while leaving it to more recent commentators to develop this line of thought in more explicitly ecological terms and to incorporate it into a broader ecological project.

One such writer is Ted Benton, who argues that 'the basic ideas of historical materialism can without distortion be regarded as a proposal for an ecological approach to the understanding of human nature and history'.[10] Benton identifies two theses of historical materialism which he believes suit it for this role: firstly, its 'naturalism', that is Marx and Engels's 'persistent view of human social life as dependent upon nature-given material conditions', and secondly, the thesis 'that the key to understanding the geographical *variations* and historical *transformations* in the form of human social and political life is to be found in the various *ways* in which these societies interact with nature'.[11] These two theses capture what I have called the existential dependence of humans upon nature and the causal influence of nature upon humanity.

Howard L. Parsons similarly attacks the notion that Marx and Engels 'denied the independent power of nonhuman nature'. Introducing a series of quotations to support his claim, he argues that

Marx and Engels repeatedly pointed to the 'objective conditions' of man's natural and historical environment within which he lives, labors, and has his being. Nature is an existence which Marx aptly described as 'presupposed' for man's communal activity. Through its instruments of labor, materials, climate, and other characteristics, nature determines man's production and in turn is determined by that production.[12]

Peter Dickens also thinks that Marxism, despite some difficulties, offers the best staging-post towards a green social theory. 'Marx and Engels', he writes, 'are arguably the only writers to have developed a science of the kind that is now needed for an adequate understanding of environmental

---

[9] In the light of such statements it is odd that many greens are drawn to post-modernism. For an account which recognises the tensions between post-modernism and environmental politics and attempts to reconcile them, see Gare 1995. Ultimately, however, Gare's account is unsuccessful: in attempting to render post-modernism compatible with the requirements of environmental politics he ends up jettisoning most of the 'radical' claims (relativism, opposition to 'metaphysics' and 'grand narratives' etc.) that are usually supposed definitive of post-modernism.   [10] Benton 1989, p. 55.

[11] *Ibid.*, p. 54. The works cited as illustration of these theses are the *Economic and Philosophical Manuscripts* ('notwithstanding the residual idealism'), *The German Ideology*, the 1859 Preface, *Capital* and the *Critique of the Gotha Programme*.

[12] Parsons 1977, p. 121. The works quoted to back up this conclusion are *The Holy Family*, *A Contribution to the Critique of Political Economy, Grundrisse, Capital*, vol. I, and the letter from Engels to Marx, 6 June, 1853. Parsons gives further quotations to illustrate other aspects of the human–nature relation in Marx: nature as dialectical, the interdependence of man and nature, their mutual transformation through labour, the effect of capitalism, and so on.

issues.'[13] Dickens expresses the dependence of social phenomena upon natural conditions in methodological terms similar to those suggested at the end of chapter 3, arguing that what is needed is to overcome the division of intellectual labour between the natural and social sciences. A consequence of this division has been that until now ecology, which is in principle applicable to humans as well as to other species, has given little recognition to the social, political, economic and gender issues which deeply affect *this* species' relation with its environment, while, on the other hand, social science has given little attention to the natural environment, and when it has been forced to address the impact of environmental problems, has focused almost exclusively on social and 'cultural' causes.[14] Themes in Marx and Engels's work identified by Dickens as offering a starting-point for overcoming the weaknesses of these one-sided approaches include people's possession of needs in common with and distinct from those of other species, and the materiality of their bodies. These factors underlie human dependence upon nature, but Dickens also notes Marx and Engels's observation of their mutual conditioning – 'in the process of changing nature, people change themselves' – as well as the ways in which this relation is mediated by social institutions and processes.

The point made by Benton, Parsons and Dickens is not simply that Marx and Engels were aware of human dependence upon nature. Such an observation would be unremarkable, even considering the historical context in which they worked,[15] and it would be of limited help in assessing the ecological significance of their account of society and its development. In order to demonstrate the ability of Marxist theory to explain the problems that arise in connection with human dependence upon nature, it is necessary to show that Marx's account of this dependence forms an integral part of his larger theory. Although Benton and Dickens (as we shall see) express reservations about Marx's consistency in this respect, all three of the writers mentioned regard the dependence of humans upon the natural environment as a central principle of Marx's work. Such a stance appears plausible, since Marx and Engels's recognition of this dependence is associated with their materialist conception of history, described by Marx himself as a 'guiding thread' for his studies.[16] We will shortly consider objections both to this 'ecological' interpretation of historical materialism,

---

[13] Dickens 1992, p. xiv.   [14] *Ibid.*, pp. 2–6.
[15] Pre-Marxian assertions of the dependence of society upon nature include Malthus's theory of population, the relation of which to Marxism is documented in chapter 2, and the claim of the Physiocrats, extensively commented upon by Marx, that the earth is the source of all value.   [16] 1859 Preface, p. 181.

and to this account of its role within the Marxist corpus. First, however, let us review some of the textual evidence in favour of such an interpretation.

The German Ideology provides perhaps the clearest account of the importance of the notion of human dependence on nature for Marx and Engels's theory of history. In this work they seek to develop their materialist approach from first principles that are elsewhere taken for granted, and to contrast their materialism with the Hegelian idealism that they had earlier espoused. At the beginning of the book they write:

> The first premise of all human history is, of course, the existence of living human individuals. Thus the first fact to be established is the physical organisation of these individuals and their consequent relation to the rest of nature. Of course, we cannot here go either into the actual physical nature of man, or into the natural conditions in which man finds himself – geological, orohydrological, climatic and so on. The writing of history must always set out from these natural bases and their modification in the course of history through the action of men.[17]

Several points may be made about this passage. The reference to the *rest of nature* indicates that Marx and Engels regard humans as a part of nature. Correspondingly, they perceive the science of history as being in some sense continuous with those sciences that investigate man's physical nature and his natural surroundings. This is a recurring theme which becomes more explicit in Marx's later works, where he compares his own theory of the development of society to Darwin's theory of natural evolution.[18] Although the objects of the natural sciences do not themselves fall within the scope of historical science as objects of investigation, they are preconditions for the existence of the human historical subject. Therefore, recognition of the dependence of human beings on their natural environment is, for Marx and Engels, essential to a proper understanding of the historical process. It is because of this dependence that Marx and Engels, a few pages later, identify production (understood as the purposive transformation of nature in order to satisfy human needs) as a fundamental and ongoing condition of history:

> we must begin by stating the first premise of all human existence and, therefore, of all history, the premise, namely, that men must be in a position to live in order to be able to 'make history'. But life involves before everything else eating and drinking, a habitation, clothing and many other things. The first historical act is thus the

[17] *The German Ideology*, p. 42.

[18] See, for example, 'Marx to Lassalle, 16 Jan 1862', p. 525: 'Darwin's book is very important and serves me as a natural-scientific basis for the class struggle in history.' Cf. Timpanaro 1975, p. 41. See also Benton 1979, in which the failure of the Left to rise to the challenges of ecological crisis is connected with the abandonment by Western Marxism of Marx and Engels's theme of the eventual unity of the sciences. The unity of science theme also appears in the *Economic and Philosophical Manuscripts*, p. 98.

production of the means to satisfy these needs, the production of material life itself. And indeed this is an historical act, a fundamental condition of all history, which today, as thousands of years ago, must daily and hourly be fulfilled merely in order to sustain human life.[19]

In his later works Marx distinguishes between the general, material characteristics of the production process, common to all its historical stages, and the specific form that it takes within a capitalist economy. In the Introduction to *Grundrisse* he cautions against treating historically specific bourgeois forms of production as 'inviolable natural laws'[20] and warns that 'the so-called *general preconditions* of all production are . . . abstract moments with which no real historical stage can be grasped'.[21] Nevertheless, he regards investigation of the general characteristics of production as a necessary step towards understanding its particular historical forms, as his plan of work indicates: 'The order obviously has to be (1) the general, abstract determinants which obtain in more or less all forms of society, but in the above-explained sense. (2) The categories which make up the inner structure of bourgeois society . . .'.[22]

In fact Marx never made a separate study of the 'general determinants' of the production process, but he did include a short discussion of the topic in chapter 7 of the first volume of *Capital*. Here he refers to the production process characterised in terms of its general, material characteristics as the *labour process*, and in its specifically capitalist form as the *valorisation process*, insisting that both are necessary to a proper understanding of the production process under capitalism.[23] Echoing the passages quoted above from the *German Ideology* Marx characterises the labour process as

purposeful activity aimed at the production of use-values. It is an appropriation of what exists in nature for the requirements of man. It is the universal condition for the metabolic interaction between man and nature, the everlasting nature-imposed condition of human existence, and it is therefore independent of every form of that existence, or rather it is common to all forms of society in which human beings live.[24]

Since the labour process involves the appropriation of non-human nature to satisfy human needs, Marx's assertion of its transhistorical necessity can

---

[19] *The German Ideology*, p. 48.   [20] *Grundrisse*, p. 87.
[21] *Ibid.*, p. 88. Cf. p. 85: '*Production in general* is an abstraction, but a rational abstraction in so far as it really brings out and fixes the common element and thus saves us repetition . . . Some determinations belong to all epochs, others to only a few. [Some] determinations will be shared by the most modern epoch and the most ancient. No production will be thinkable without them . . .'.   [22] *Ibid.*, p. 108.
[23] 'Just as the commodity itself is a unity formed of use-value and value, so the process of production must be a unity, composed of the labour process and the process of creating value.' *Capital*, vol. I, p. 293.   [24] *Ibid.*, p. 290.

be seen to imply the existential dependence of humans upon non-human nature (the principle of existential dependence).

Later in this chapter we shall see how Marx's 1859 Preface has been used to support a different interpretation of historical materialism from that suggested by the preceding material. In the full work from which the Preface is taken, however, we find the following passage:

> It would be wrong to say that labour which produces use-values is the *only* source of the wealth produced by it, that is of material wealth. Since labour is an activity which adapts materials for some purpose or other, it needs material as a prerequisite. Different use-values contain very different proportions of labour and natural products, but use-value always comprises a natural element. As useful activity directed to the appropriation of natural factors in one form or another, labour is a natural condition of human existence, a condition of material interchange between man and nature, quite independent of the form of society.[25]

Here, Marx again asserts both the transhistorical dependence of humans upon the labour process and the essential part that natural 'elements' or 'factors' play in that process. The latter point is emphasised once more in the *Critique of the Gotha Programme* where, responding to the claim that labour is the source of all wealth, Marx replies that '[n]ature is just as much the source of use-values (and it is surely of such that material wealth consists!) as labour, which itself is only the manifestation of a force of nature, human labour power'.[26]

The quotations given so far are drawn from the *German Ideology* and subsequent works. Marx's *Economic and Philosophical Manuscripts*, written a year or so earlier under the influence of Feuerbach, are not considered part of Marx's mature work. They have, however, been regarded by some writers as the most ecological of Marx's works. In the next section I shall be examining differing views on the periodisation of Marx's work, as applied to ecological issues; for now, however, it can be noted that the conception of the human species as a dependent part of nature, which we have found in Marx's and Engels's later works, is prefigured in a widely quoted passage from the *Economic and Philosophical Manuscripts*: 'Nature is man's *inorganic body* – nature, that is, insofar as it is not itself human body. Man *lives* on nature – means that nature is his *body*, with which he must remain in continuous interchange if he is not to die. That man's physical and spir-

---

[25] *A Contribution to the Critique of Political Economy*, p. 36.

[26] *Critique of the Gotha Programme*, p. 315. Here Marx is reiterating a point made in *Capital*, vol. I, p. 638, where he states that capitalist production only develops 'by simultaneously undermining the original sources of all wealth – the soil and the worker'.

itual life is linked to nature means simply that nature is linked to itself, for man is a part of nature.'[27]

The passages quoted above, from some of Marx's most important works, appear to commit him unequivocally to the principle of existential dependence. In the *Economic and Philosophical Manuscripts* and the *German Ideology* this is expressed in terms suggestive of what I have called the principle of ecological inclusion. The later passages, from *A Contribution to the Critique of Political Economy*, *Grundrisse* and *Capital*, continue to assert the dependence of humans upon nature, but do not express this in the language of inclusion. However, this would appear to be a matter of presentation and emphasis (corresponding perhaps to a shift from general statements of the human–nature relation to a more concrete analysis of particular interactions between humans and non-human nature), rather than a fundamental change of outlook, since the notion of humans as a part of nature reappears in the quotation from the *Critique of the Gotha Programme*. The fact that Marx considered knowledge of the material prerequisites of human existence to be necessary for an understanding of society suggests additionally a recognition of the causal influence that nature exercises over human societies even once the conditions of human existence are satisfied (the principle of causal dependence). Critics have argued, however, that Marx's commitment to the principle of ecological dependence is neither as consistent, nor as central to his work, as the quoted passages suggest, and it is to these objections that I now turn.

4.3 Early and late Marx: an ecological break?

It is often felt by environmental commentators on Marx that whatever insight he had concerning the relation between humans and external nature is concentrated in his early writings and was later abandoned or forgotten.

Fritjof Capra, for example, quotes approvingly the passage from the *Economic and Philosophical Manuscripts* in which Marx describes nature as man's inorganic body. He also quotes from the same work (though, as illustrated above, he could equally well have quoted from one of Marx's later works) to demonstrate that for Marx nature as much as labour is

---

[27] *Economic and Philosophical Manuscripts*, p. 67. Marx also writes of nature as the worker's 'means of life' (*ibid.*, p. 64, quoted in chapter 6 below, text to note 71), and of man as a natural, corporeal and conditioned being with needs for things that exist outside and independent of him (p. 136). See also p. 98 on the eventual unity of natural science and the 'science of man'.

necessary to the productive process.[28] Capra does offer a quotation from *Capital* to illustrate Marx's awareness of the ecological impact of capitalist economics, but generally he is critical of the striving for scientificity in Marx's later economic works and regards Marx's ecological insights as being *despite* what he perceives as the 'technological determinism, which made [Marx's] theory more acceptable as a science'.[29]

Similarly, Donald C. Lee seeks 'to direct [orthodox] Marxists toward a position superior to the outmoded nineteenth-century world view they are caught in',[30] by arguing for a humanist Marxism premised upon the centrality of Marx's pre-1845 works. '[B]oth Marxism and capitalism', he writes, 'are greedy, violent, and destructive of nature *unless* they are ameliorated by that humanistic view.'[31]

Even Peter Dickens who, as we saw above, considers Marxism to be the best approximation yet to an ecological social theory, endorses a more attenuated form of this periodisation. Although Dickens does not say that Marx and Engels ever abandoned the ecological themes that he identifies in their earlier works, and maintains that some of these themes – notably alienation and fetishisation – continued to be used in their later work, he argues that the later works failed to develop their early ideas on the relationship between humans and nature.[32] A similar position has been taken by Rudolph Bahro:

For the young Marx, the abolition of private property in the means of production was precisely to bring about the reconciliation of culture and nature. Marx already perceived the contradiction between capitalist production and nature. It was just that this was as yet not so acute for him to place it at the centre of his analysis. Later on this point of view retreated more to the margin of Marx's thinking, since he concentrated closely on analysing the capital relationship in the stricter sense, i.e. on the problems that at that time stood in the way of the further development of working humanity.[33]

The question of continuity and change generally in Marx's thought is too large and complex to be considered here. As regards Marx's ecological views, we may concede that certain ecologically significant themes are more prominent in Marx's early works. This will be apparent in the discussion of Marx's conception of human needs in chapter 6. On the more specific issue of Marx's adherence to the principle of ecological dependence, we have already seen, in the previous section, that the conception of humanity as a dependent part of nature was present in Marx's major works throughout the period of his mature writings, from his early elab-

---

[28] Capra 1983, p. 216.   [29] *Ibid.*, pp. 218–19.   [30] Lee 1982, p. 341.
[31] Lee 1980, p. 4.   [32] Dickens 1992, p. xiv.   [33] Bahro 1982, p. 30.

oration of historical materialism in 1846, through his economic works of the 1850s and 60s, and as late as 1875. The suggestion that Marx abandoned this conception of the human–nature relation is therefore implausible, though it remains to be seen what role this conception plays in the overall structure of Marx's thought.

Some writers have adopted the opposite stance, arguing that it is *only* in Marx's later works that a proper recognition of the role of nature is to be found. Timpanaro, for example, writes that 'Marxism was *born* as an affirmation of the decisive primacy of the socio-economic level over juridical, political and cultural phenomena, and as an affirmation of the historicity of the economy.'[34] The materialism of the early Marx, in other words, was simply a denial of the Hegelian tradition in the philosophy of history, which affirmed 'the primacy of the spirit over any economic structure'. By contrast, 'Marx in his maturity – who admired Darwin and wanted to dedicate the second volume of *Capital* to him, who declared in the preface to *Capital* itself that he "viewed the evolution of society as a process of natural history" – was certainly much more materialist than the Marx of the *Theses on Feuerbach.*'[35]

It is certainly plausible to maintain that Marx's views became more materialist as he moved further from his early identification with Hegelianism. Given this genesis, and the evidence presented above for Marx's commitment to the principle of ecological dependence in his mature works, a periodisation in the manner of Timpanaro may seem more plausible than that suggested by Capra and others. If we confined our interest to Marx's mature theory we could accept Timpanaro's view and maintain only that Marx's later works embraced the principle of ecological dependence. However, I do not believe that the contrast between the earlier and later Marx is that stark. In chapter 6 I will show how the theory of human nature and human need developed in Marx's early works is relevant in assessing the ecological implications of his mature theory. For now, however, it is sufficient to note that the passage from the *Economic and Philosophical Manuscripts* quoted above demonstrates a recognition by Marx of the dependence of humanity upon nature as integral to his enterprise of understanding society, even before the alleged 'break' in his intellectual development. The terms in which he expresses this dependence are also indicative of a continuity with the account given in his later works.

Given this continuity, the question arises of why it is suggested that Marx neglected or abandoned the principle of ecological dependence in

---

[34] Timpanaro 1975, p. 40; emphasis changed.   [35] *Ibid.*, p. 41.

either his earlier or his later works. The answer, I believe, is to be found in other features of these works, which are perceived by many of Marx's environmental critics as contrary to the principle of ecological dependence, and which therefore lead them to overlook or misinterpret the statements implying this principle to be found in these works. Among the more significant of these supposedly anti-ecological features of Marx's *later* works are the notion of the development of the productive forces (development of human capacity to transform nature), and, connected with it, the notion of a succession of economic forms and the explanatory role accorded to these forms in relation to ecological and other problems. We saw above that the transformation of nature by humans does not necessarily contradict the dependence of humans upon nature, and is in fact an essential component of an ecological view of society. Judgement must therefore be reserved on Marx's account of the development of the productive forces, and I will return to this in the following chapters. Similarly, it will be seen in the following section that those who see in Marx's *early* works a neglect of human dependence upon nature do so on the basis of an interpretation of what Marx says in those works about transformation of nature by humans, which they interpret as being incompatible with that dependence.

4.4 Developing the active side: idealist interpretations of historical materialism

Marx and Engels regard their own materialism as differing importantly from earlier variants. The key distinguishing feature, they maintain, is the attention that they give to the *transformation* of nature by humans:

> In reality and for the *practical* materialist, i.e. the *communist*, it is a question of revolutionising the existing world, of practically attacking and changing existing things. When occasionally we find such views in Feuerbach, they are never more than isolated surmises and have much too little influence on his general outlook to be considered here as anything else than embryos capable of development . . . He does not see how the sensuous world around him is, not a thing given direct from all eternity, remaining ever the same, but the product of industry and of the state of society . . .[36]

In criticising Feuerbach's 'contemplative' materialism for its neglect of human activity, Marx sees himself as appropriating for materialism the *active* side' which, he argues, had hitherto been 'developed by idealism'.[37] Some writers, however, have seized upon this emphasis on human activity in order to construct an interpretation of Marxism that is hardly recog-

---

[36] *The German Ideology*, p. 62.   [37] *Theses on Feuerbach*, p. 28.

nisable as materialism at all, and appears clearly opposed to the principle of ecological dependence.

It is towards writers such as these that Timpanaro's defence of materialism is directed. Timpanaro labels such writers 'Left-idealists', and notes that though theirs may be a disguised idealism in which the social construction of nature is expressed in terms of 'praxis' rather than 'thought', the change in terminology makes little real difference; whichever term is used, the existence of nature independent of humanity is denied, and it is this that Timpanaro objects to. Left-idealists, he argues, cannot recognise that the fate of conscious beings may depend upon the external world, since they hold that the external world itself is no more than the contents of their thought.[38]

A writer who clearly exemplifies this viewpoint is Leszek Kolakowski. He denies that Marx has any conception of either human nature or external nature that is not socially constituted, and complains of Marx's 'lack of interest in the natural (as opposed to economic) conditions of human existence, the absence of corporeal human existence in his vision of the world. Man is wholly defined in purely social terms; the physical limitations of his being are scarcely noticed.'[39]

This assertion is puzzling in the light of the passages quoted earlier, in which Marx highlights the very issues that Kolakowski accuses him of ignoring. The explanation would appear to be as suggested by Timpanaro: that Kolakowski interprets the natural or material world to which Marx refers in those passages as wholly a product of human consciousness. Thus Kolakowski holds that for Marx the existence of things 'comes into being simultaneously with their appearance as a picture in the human mind',[40] and that 'Marx's view was that nature as we know it is an extension of man, an organ of practical activity'.[41] But while these views clearly have their origins in passages like that quoted at the beginning of this section, in which Marx emphasises the transformative power of human agency, nothing in these passages implies the denial of a reality independent of human consciousness asserted by Kolakowski.

A more natural reading, and one which is more readily reconcilable with Marx's assertions of human dependence on nature, would involve, first, the idea that there is a pre-existing material reality upon which humans depend, which conditions their consciousness and constrains their actions. This forms the material 'premise' of human existence, ignored, as Marx

---

[38] Timpanaro 1975, p. 101.   [39] Kolakowski 1978, pp. 412–13.
[40] Kolakowski 1969, p. 69.   [41] Kolakowski 1978, p. 401.

says, by German idealist philosophy.[42] Second, that human activity, conditioned and constrained as it is by that reality, nevertheless changes it in profound ways, so that the conditions in which humans find themselves at any time are not 'given direct from all eternity' but are 'an historical product, the result of the activity of a whole succession of generations each standing on the shoulders of the previous one'.[43] And third, that human agency nevertheless remains conditioned and constrained by the circumstances in which it operates, and dependent upon that external reality; hence Marx's characterisation of the appropriation of that external reality as a 'fundamental' or 'universal' condition of all history.[44] It is true that in his comments on Feuerbach[45] Marx highlights the extent of human transformation of nature rather than its preconditions or limits, but this is to counter what he sees as the weakness of Feuerbachian materialism: its tendency to see humans as passive products of their circumstances and to forget 'that circumstances are changed by men'.[46] Elsewhere, as we have seen, Marx chooses to highlight the other side of the relation.

Another commentator labelled a 'Left-idealist' by Timpanaro is Alfred Schmidt, though in fact Schmidt appears to equivocate between Kolakowski's idealist interpretation of Marx and the more conventional one proposed above.[47] A detailed catalogue of Schmidt's 'idealist jargon' is given by David-Hillel Ruben, so a few examples will suffice here for illustration.[48] Among the most obvious of these are the statements that 'material reality is *from the beginning* socially mediated', that '[n]atural history is human history's extension backwards', and that 'matter must appear as a social category in the broadest sense'.[49] Phrases such as these would appear to warrant Timpanaro's label; however, Timpanaro ignores other aspects of Schmidt's writing which suggest a different interpretation.

One of Schmidt's aims is similar to the one I have been pursuing: to show that, for Marx, human activity is and will remain dependent upon nature. Thus he argues that for Marx 'the social mediation of nature confirms its "priority" rather than abolishes it'; 'man's aims can be realized by the use of natural processes, not despite the laws of nature but precisely

---

[42] See text to notes 17 and 19 above.   [43] *The German Ideology*, p. 62.
[44] See text to notes 19 and 24 above.
[45] In the *Theses on Feuerbach* and *The German Ideology*, pp. 60–4.
[46] *Theses on Feuerbach*, p. 28. Marx's comment in the *German Ideology*, p. 59, 'that circumstances make men just as much as men make circumstances' makes essentially the same point but with the emphasis reversed. The co-existence of agency and conditioning is also expressed in another context in Marx's aphorism that men make their own history but not in circumstances chosen by themselves (*The Eighteenth Brumaire of Louis Bonaparte*, p. 96).
[47] Schmidt 1971.   [48] Ruben 1979, pp. 83–5.
[49] Schmidt 1971, pp. 35, 46, 32; emphasis added.

because the materials of nature have their own laws', and conversely the aims that can be realised are limited not just 'by history and society but equally by the possibilities immanent in matter itself'.[50] Marx's material-ism, he explains, does not deny 'that matter has its own laws and its own movement', but understands 'that matter's laws of motion can only be rec-ognized and appropriately applied by men through the agency of mediat-ing practice'.[51] Marx, Schmidt argues, came to the view 'that the struggle of man with nature could be transformed but not abolished', and thus 'rec-onciled freedom and necessity on the basis of necessity'; and finally, he 'insisted at many different points in *Capital* that labour could never be abol-ished'.[52] All of this is consistent with the 'materialist' interpretation of Marx's allegedly 'idealist' passages proposed above, and with the princi-ple of ecological dependence.

But if the main tendency of Schmidt's writing is materialist in a sense consistent with the principle of ecological dependence, what is to be made of his idealist language? Ruben points out that phrases like those used by Schmidt may, in addition to their obvious idealist sense, be interpreted in a weaker, perhaps trivial, sense. Schmidt's claim that '[n]atural history is human history's extension backwards', for example, may be interpreted trivially as the assertion that 'it is impossible to *speak* of natural history unless one presupposes human history, for without human beings there could be no speaking about anything'.[53] There is also, Ruben notes, a 'broad and obvious sense in which any category is social'. If this is what Schmidt means by saying that 'matter must appear as a social category in the broadest sense', this is a 'true but rather dull' claim.[54] These weaker interpretations may be described as epistemological, rather than ontolog-ical, uses of traditionally idealist language, since the subjective element alluded to is asserted as an aspect of our thought or communication about the world, not of the external world itself. But while at least some of Schmidt's 'idealist' phrases may be defended in this way, the overall manner of their presentation is at best obfuscatory, suggesting as it does an interpretation of Marx that is at odds with the 'ecological' interpretation argued for here as the most plausible reading of his texts and which Schmidt himself appears to accept.

---

[50] *Ibid.*, pp. 96, 63.

[51] *Ibid.*, p. 97. In support of this Schmidt quotes Marx's comment (in his letter to Kugelmann, 11 July 1868), that '[i]t is absolutely impossible to transcend the laws of nature. What *can* change in historically different circumstances is only the *form* in which these laws express themselves.' The reference to a change in the form in which the laws express themselves presumably refers to changes in the way they are used by humans.

[52] *Ibid.*, pp. 76, 136. See also p. 71.   [53] Ruben 1979, p. 85.   [54] *Ibid.*, p. 84.

4.5 Productivism and the labour process: Benton's critique

The portrayal of Marx as an idealist by commentators like Kolakowski and (less clearly) Schmidt contrasts with Ted Benton's aforementioned view of historical materialism as an acknowledgement of humans' dependence upon nature. As noted earlier, however, Benton himself questions Marx's consistency in applying this materialist perspective. According to Benton there is a hiatus in Marx and Engels, not between their earlier and later works, but between their 'materialist premises in philosophy and the theory of history' and an economic theory in which 'there is a significant retreat from [this] thoroughgoing materialism'.[55] At the core of Benton's critique is a claim about the inability of Marx's basic economic categories to articulate the full extent to which economic processes depend upon ecological factors.

This is a somewhat surprising claim, since the categories in question are those set out in chapter 7 of *Capital*, where as we saw earlier Marx characterises the production process, conceived in its material aspect as the 'appropriation of what exists in nature for the requirements of man', as 'the everlasting, nature-imposed condition of human existence'.[56] Benton argues, however, that the key concepts in which this account of the labour process is framed 'involve a series of related conflations, imprecisions and lacunae, the net effect of which is to render the theory incapable of adequately conceptualizing the ecological conditions and limits of human need-meeting interactions with nature'.[57]

Marx's account of the labour process involves a primary classification into three elements: '(1) purposeful activity, that is work itself, (2) the object on which that work is performed, and (3) the instruments of that work'.[58] *Objects of labour* are further divided into those he terms 'raw materials', which are the result of previous labour processes,[59] and those which are 'spontaneously provided by nature'. Raw materials are further subdivided into those which form the 'principal substance' of a product, and 'accessories'. The latter include materials such as fuels and lubricants consumed by the instruments of production, substances such as dyes or coal added to iron, which are added to the raw material 'in order to produce some physical modification of it', and materials used 'to help accomplish the work', for example in the lighting and heating of a workshop.[60]

---

[55] Benton 1989, p. 55.   [56] See text to note 24 above.   [57] Benton 1989, p. 63.
[58] *Capital*, vol. I, p. 284.
[59] Thus Marx uses 'raw materials' in a sense that is narrower than that of standard English, and somewhat counter-intuitive.   [60] *Capital*, vol. I, p. 288.

The *instruments of labour*, like its objects, include both man-made and natural items. Marx defines an instrument of labour as 'a thing, or a complex of things, which the worker interposes between himself and the object of his labour and which serves as a conductor, directing his activity onto that object'.[61] In the simplest cases of labour activity, such as fruit gathering, man's limbs may be considered as instruments of labour. Stones and other natural objects function as instruments of labour at an early stage. 'As soon as the labour process has undergone the slightest development', however, 'it requires specially prepared instruments.' The earth, Marx holds, functions as an instrument of production in agriculture, but only as the result of its modification by previous labour: 'its use in this way . . . presupposes a whole series of other instruments and a comparatively high stage of development of labour-power'. Marx also includes as instruments of production 'in a wider sense' things which do not serve directly as 'conductors' of man's activity but are 'the objective conditions necessary for carrying on the labour process'.[62] Like instruments in the narrower sense, these can be natural or man-made: 'the earth itself is a universal instrument of this kind, for it provides the worker with the ground beneath his feet and a "field of employment" for his own particular process. Instruments of this kind which have already been mediated through past labour, include workshops, canals, roads, etc.'[63]

Benton's critique of Marx's conception of the labour process comprises two distinct claims. First, he argues that it fails in its claim to universality, since it assumes an 'intentional structure' applicable only to one broad type of labour process which he terms 'productive'[64] or 'transformative'. Other types of labour process, namely 'eco-regulation' (agriculture and similar processes) and 'primary appropriation', have features which make them particularly dependent upon environmental conditions, and which cannot, he claims, be assimilated by the categories of the 'productive' labour process. Benton's second claim is that even 'productive' labour processes have features that are important environmentally but are excluded from Marx's account.

Benton lists four features of eco-regulation and two of primary appropriation which, he claims, are not shared by 'productive' labour activities, and cannot be accounted for by Marx's concept of the labour process.[65]

---

[61] *Ibid.*, p. 285.   [62] *Ibid.*, p. 286.   [63] *Ibid.*, pp. 286–7.
[64] Benton thus uses the term 'productive' in a narrower sense than Marx, which I will indicate by the use of quotation marks. Marx defines productive labour, from the standpoint of the labour process, as labour which produces use-values (*Capital*, vol. I, p. 287). This can include processes of 'eco-regulation' and 'primary appropriation' in contrast to which Benton's concept of productive labour is defined.   [65] Benton 1989, pp. 67–8.

(1) In eco-regulation labour is applied to the 'conditions' of labour, rather than to the raw materials. (Benton uses 'conditions' or 'contextual conditions' to refer to what Marx calls instruments of labour in a wider sense – i.e. those which are necessary conditions of the labour process but do not directly 'conduct' man's activity to its object.) (2) This labour sustains, regulates and reproduces rather than transforming the substance upon which it operates (the soil, for example). (3) The spatial and temporal distributions of this labouring activity are strongly shaped by its contextual conditions and 'by the rhythms of organic development processes'. (4) Finally, nature-given conditions 'figure both as *conditions* of the labour-process, *and* as [*objects*] of labour' thus forming a category 'not readily assimilable to Marx's tripartite classification'.[66]

My contention, however, is that these features are not unique to eco-regulation and may be assimilated within Marx's conceptual scheme. Benton's assertion that in eco-regulation labour is not applied primarily to the raw materials is misleading since it is usual in all kinds of labour process for labour to be applied directly not to the raw materials but to the instruments of production. More specifically, it is common not only in agriculture but also in industrial production for labour to be applied to what Benton calls the contextual conditions and Marx calls instruments of production in the wider sense: buildings, infrastructure, etc. Here too we find labour directed immediately at maintaining or sustaining rather than transforming. (Of course the 'contextual conditions' to be sustained in industrial production may be perceived as 'less natural' than those involved in agriculture, but agricultural systems themselves involve substantial modification of the natural systems from which they derive, as Marx asserts and Benton acknowledges.)[67]

Similarly, the dependence of the location and timing of the labour process upon natural conditions may be more pronounced in the case of agriculture than in manufacturing, but it is present in the latter too. The location of a factory, for example, will depend amongst other things upon the proximity of raw materials;[68] and if those raw materials are seasonal (as is the case where the product of an agricultural labour process forms the raw material for a 'productive' labour process) it is likely that the

[66] In (4) I have changed Benton's 'subjects' to 'objects' of labour to render it consistent with the translation of Marx used in the preceding exposition of his concept of the labour process. Benton's terminology comes from the London, 1961 edition of *Capital*, and is liable to be misleading since, on one interpretation of 'subject', the subject of labour would be the person doing the labouring. Cf. Cohen 1978, p. 38.

[67] *Capital*, vol. I, p. 285 (text to note 61 above); Benton 1989, pp. 66–7; Benton 1992, p. 61.

[68] Here, and in what follows, I am using 'raw materials' in its everyday English sense, equivalent to Marx's 'objects of labour'. (Cf. note 59 above.)

scheduling of production will be affected.[69] 'Productive' labour processes may also be directly affected by climatic conditions to the extent of being made impossible, or uneconomically expensive, at certain times and/or in certain locations. To the extent that labour processes of any type are dependent upon climatic or other natural conditions, those conditions fall within Marx's classification as instruments of labour in the wider sense.

Finally, the fact that nature-given conditions can serve both as conditions and as objects of labour is explicitly allowed by Marx in his account of the labour process. Since '[e]very object possesses various properties, and is thus capable of being applied to different uses', he writes, it is possible for the same object to 'be used as both instrument of labour and raw material in the same process'.[70]

The features that Benton identifies as typical of primary labour processes are susceptible to a similar critique. Firstly, Benton argues that in this type of process the conversion of a natural material into a use-value cannot be adequately described as 'Nature's material adapted by a *change of form* to the wants of man'.[71] This depends upon a narrow interpretation of 'change of form' and Benton gives no grounds for his assumption that Marx's use of this phrase excludes 'selecting, extracting and relocating elements of the natural environment'. Indeed, these processes, along with synthesis of various of the selected and extracted parts, are all elements of 'productive' labour processes. Secondly, Benton argues that primary labour processes share with eco-regulation a dependence upon naturally given external conditions. We have seen, however, that this is a dependence which is shared in varying degrees by all types of labour process.

Benton writes that 'the extent of the difference between eco-regulatory and productive practices is the measure of the inadequacy of Marx's abstract concept of the labour process'.[72] The foregoing argument indicates, however, that the difference is not as sharp as Benton suggests. The features cited by Benton as characteristic of eco-regulation and primary appropriation are shared in varying degrees by all types of labour process, and may be captured in the terms of Marx's account. Benton might reply that this overlooks differences in the *degree* to which different types of labour process are dependent upon natural conditions. Such an objection, however, would ask too much of a general concept. A general concept, by its nature, does not specify the differences between the objects to which it

---

[69] This point is made in *Capital*, vol. III, p. 213.   [70] *Capital*, vol. I, p. 288.
[71] Benton 1989, p. 69, quoting (with emphasis added) from the London 1961 edition of *Capital*, vol. I. The equivalent phrase in the Penguin/New Left Review edition cited here is on p. 287.   [72] *Ibid.*, p. 67.

applies, although neither does it imply that such differences are insignificant. Marx does not claim that his concept of the labour process 'in its simple and abstract elements' constitutes an exhaustive description of the material characteristics of the labour process, or one that is sufficient for all purposes; indeed, he emphasises in *Grundrisse* that such a concept abstracts not only across historical epochs but also across branches of production, picking out the 'common element' but ignoring the differences that characterise the particular cases.[73]

The second part of Benton's critique is that even with respect to 'productive' labour processes, Marx's account of the labour process is inadequate. Three corrections, he claims, are required:

First, contextual conditions should be conceived separately from the instruments of labour, as an independent class of 'initial conditions'. Second, the continuing pertinence of these contextual conditions to the *sustainability* of production needs to be incorporated, as with eco-regulatory practices. This is significant in that it renders thinkable the possibility that these conditions might cease to be spontaneously satisfied, and so require the ancillary labour-process of restoring or maintaining the environmental conditions for productive sustainability . . . Third, some of the naturally mediated unintended consequences of the operation of labour-processes may impinge upon the persistence or reproduction of its contextual conditions and/or raw materials.[74]

Clearly from an environmental perspective it is necessary to examine the role of what Benton calls 'contextual conditions' in human activity. To apply a separate term to this element of the labour process, as Benton does, may therefore be convenient. But the distinction between these instruments of labour 'in a wider sense' and the instruments of labour more narrowly defined is already made by Marx. In emphasising that these conditions 'cannot plausibly be considered "conductors" of the activity of the labourer', Benton is echoing Marx's own account of this distinction.[75]

The possibility that these conditions may cease to be satisfied and the consequences of this for the sustainability of production must also be recognised. But there is nothing in Marx's concept of the labour process which precludes such a possibility, and indeed observations that Marx and Engels make elsewhere show that they recognise this, not just as a possibility but as a reality. In *Capital*, for example, Marx writes that capitalist production, because of the way it concentrates the population in urban centres,

---

[73] *Grundrisse*, pp. 85–6. See also text to note 21 above.   [74] Benton 1989, p. 73.
[75] *Ibid.*, p. 72; cf. *Capital*, vol. I, p. 286.

disturbs the metabolic interaction between man and the earth, i.e. it prevents the return to the soil of its constituent elements consumed by man in the form of food and clothing; hence it hinders the operation of the eternal natural condition for the lasting fertility of the soil . . . [A]ll progress in capitalist agriculture is a progress in the art, not only of robbing the worker, but of robbing the soil; all progress in increasing the fertility of the soil for a given time is a progress towards ruining the more long-lasting sources of that fertility.[76]

Engels, in a passage quoted above, expresses a similar worry about the unintended effects that productive activity may have upon its own natural basis.[77] He warns that we should not 'flatter ourselves overmuch on account of our human victories over nature. For each such victory nature takes its revenge on us. Each victory, it is true, in the first place brings about the results we expected, but in the second and third places it has quite different, unforeseen effects which only too often cancel the first.'[78] Marx and Engels thus recognise not only that the contextual conditions of human production may cease to be satisfied, but also that this can come about through the unintended consequences of productive activity itself. The category of unintended consequences may therefore be appended to Marx's account of the labour process without violating his or Engels's beliefs. Marx's and Engels's examples also illustrate the two senses in which a consequence of human action can be unintended: it can be unforeseen by the agent owing to a lack of knowledge, or it can simply be outside the range of considerations included in the agent's calculations. Either way, it is something that happens in addition to (or possibly instead of) the predicted outcome that motivates the agent's productive activity. Benton explains such happenings as results of the 'residual' properties possessed by components of the labour process over and above those that enable them to perform their intended function.[79] This is a useful way of conceptualising a certain kind of ecological problem, but one which can readily be incorporated within Marx's account of the labour process as a simple extrapolation of his observation that '[e]very object possesses various properties, and is thus capable of being applied to different uses'.[80]

In conclusion, Benton has shown not that Marx's concept of the labour process renders environmental problems 'unthinkable', or that it marks a retreat from his recognition of human dependence on nature, but only that minor revisions are necessary in order to bring into the foreground points

---

[76] *Capital*, vol. I, pp. 637–8. See also vol. III, pp. 949–50, where Marx writes of the 'squandering', 'exploitation', 'laying waste', 'ruining' and 'exhausting' of the earth's natural power and vitality.   [77] See text to note 63, ch. 3.
[78] 'The Part Played by Labour in the Transition from Ape to Man', pp. 361–2.
[79] Benton 1989, p. 73.   [80] *Capital*, vol. I, p. 288.

that are already implicit either in the account of the labour process itself or elsewhere in Marx and Engels's writings.

4.6 Productivism and historical materialism: Blackburn's critique

Another commentator who, like Benton, criticises Marx for being too narrowly focused on productive activity and insufficiently thoroughgoing in his materialism is Richard James Blackburn. For Blackburn, however, these shortcomings lie not in the detailed working-out of Marx's theory but in its basic explanatory orientation. Marx, according to Blackburn, offers a theory of history which only partially lives up to its 'materialist' billing. His emphasis on the role of material production, important as it is in explaining social change, is coupled with a neglect of other factors which have an equal or greater claim to be included within a materialist explanation:

Marxist productivism has involved an emphasis upon the development of material, rather than cultural, production as the explanans of historical changes in the last instance. It is not clear why an ultimately economic explanation of this type should be considered as more materialist than a military one that focuses on problems of material destruction and its avoidance, or why either should be more materialist than a geographical or even demographical one.[81]

The factors missing from Marx's account, according to Blackburn, are broadly those addressed within the competing paradigm of geopolitics.[82] Geopolitics, however, focusing on the way in which societies are shaped by their relations with other societies and their geographically and historically variable natural environments, neglects the internal dynamics of property systems, class structure etc., which also influence their development.[83] What is therefore needed, Blackburn proposes, is a synthesis of Marxism and geopolitics.[84]

Blackburn sees Marx's 'productivism' as symptomatic of a more general narrowness of perspective, shared with Hegel. History, according to Blackburn, involves a dialectic of creation and destruction, but Marx and Hegel overemphasise creation and recognise destruction only as a secondary phenomenon arising from contradictions within the constructive moments themselves.[85] A properly materialist theory, by contrast, would attend also to the destructive forces that act on a society from outside.

---

[81] Blackburn 1990, p. 13.
[82] *Ibid.*, p. 1. Blackburn defines geopolitics as 'a study of the direct influence of historical geography, geoeconomics and historical demography upon the politics and especially the foreign relations of a state and of the influence in time of the latter activities upon the former.' (*Ibid.*, p. 9.)   [83] *Ibid.*, p. 12.   [84] *Ibid.*, ch. 3.   [85] *Ibid.*, pp. 2, 13.

These include the destruction wreaked by nature, including natural disasters such as earthquakes, slower processes of desertification and the like, disease epidemics, and the slow but incessant process of entropic decay, and also the destruction wreaked by one society on another, in the form of war, conquest, etc. A materialist theory should therefore recognise that resistance to such destructive forces – by means which include military capability, medicine and ecological control – is as much a precondition for a society's survival as the production of material necessities emphasised by Marx.[86]

As an ecological critique, Blackburn's account might be thought misdirected, in that many of the ecological problems faced by humans do arise from within their own creative engagement with nature, as unintended consequences of their productive activity, and not as raw consequences of nature's destructiveness.[87] Even so-called 'natural disasters', such as floods and hurricanes, are increasingly associated with human activity through the mechanism of global warming. Nevertheless, Blackburn's highlighting of the destructive forces of nature may be welcomed, since there are cases enough where nature wreaks destruction on human societies without any human assistance, and even where human activity does lead to ecological problems this is often a matter of encouraging or unleashing forces that already exist, rather than creating them *ex nihilo*.[88] The question, however, is whether Marx's materialism is as blind to this destructive side of nature as Blackburn suggests.

In the passage quoted at the beginning of the present section, Blackburn suggests that an explanation of historical development in terms of production is no more materialist than one which focuses on military capacity, and less so than one which focuses on demography or geography. The point about militarism, for Blackburn, is that production is insufficient to satisfy material needs since, in the face of threats from other societies, it is also necessary to safeguard what has been produced. Marx, however, has plausible reasons for considering production more fundamental, first, because while militarism may have been an influential force for much of

---

[86] *Ibid.*, pp. 2–3.
[87] It will be recalled, however, that in chapter 1 I avoided making it a definitional requirement of ecological problems that they be caused by human agency.
[88] Earthquakes and volcanic eruptions provide examples of naturally destructive forces in which human intervention typically plays no causative role. Another (potential) example much discussed recently is the effect of a meteorite impact on Earth. Hurricanes and floods provide examples of natural phenomena which have always impacted destructively on humans but which may be becoming more frequent owing to emissions of greenhouse gases. Similarly, infectious disease is a natural phenomenon, but one which may be encouraged by concentration of population, mobility and poor sanitation.

human history it is not (as production is) a necessary condition of *all possible* societies, and second, because the significance and consequences of military action will depend upon the nature of the productive forces in the societies involved more than vice versa. What the latter suggests is that production, in Marx, should be seen not only as a creative or constructive enterprise, but also as a means of combating one of the destructive external forces highlighted by Blackburn.

The issue of militarism (or more generally the destructive impact of other societies) may appear to be a side-issue here, but the same point may be made in relation to the destructive impact of nature. Blackburn criticises Marx's focus on *scarcity* for conceiving nature's adversity too passively – in terms of its failure to provide the means for humans to satisfy their needs as generously as they might like, rather than in terms of its actively destructive power.[89] However, a response to this is suggested by Blackburn's own observation that material needs themselves arise from the struggle of biological organisms to maintain themselves in the face of the disintegrative forces of entropic nature.[90] If the needs for food, clothing, shelter and so on are conceived as consequences of nature's destructiveness, then an acknowledgement of that destructiveness is already implicit within the notion of scarcity; and production, viewed by Marx as a response to that scarcity, can be viewed not only as a constructive act but also as an attempt to pre-empt or rectify the effects of those destructive powers.[91] Blackburn's emphasis on the destructiveness of nature should not, then, be seen as contradicting Marx's focus on production but rather as elaborating (more fully than Marx did) the conditions that render productive activity necessary – the conditions that Marx, in the *German Ideology*, described as 'natural bases' for the writing of history, though outside the scope of that enquiry.[92]

Blackburn also writes that geography and demography are more central to a materialist account of history than either economic production or militarism, since they 'refer to forms of matter which are of fundamental relevance for any materialist theory of history – the local configurations of inanimate nature and of animal and human populations of it'.[93] The idea that 'forms of matter', or 'configurations of nature', lie at the heart of a

---

[89] Blackburn 1990, pp. 16–17.   [90] *Ibid.*, p. 18.
[91] Marx's emphasis on *creativity* in characterising human productive activity is not intended to deny that such activity is in a sense a *reaction* to facts about their natural surroundings; rather, it is intended to highlight differences in the *ways* in which humans and other animals react to natural adversity – differences which, as we shall see, are important for Marx in explaining why humans, unlike other animals, have a social as well as a natural history.   [92] See text to note 17 above.   [93] Blackburn 1990, pp. 13–14.

materialist explanation may, again, plausibly be assimilated to Marx's idea that the physical nature of humans and their environment constitute 'natural bases' for historical enquiry. But, as we saw in chapter 2, Marx argues that human demography is socially conditioned and therefore cannot be treated, as it is by Malthus, as a purely natural determinant of social development. Animal populations too, like other elements of the environment, are altered by human action. It is therefore not clear why human or animal demography (as opposed to, say, their biology) should be considered more fundamental than production in a materialist explanation of history.

This leaves geography. A geographical explanation of history would presumably focus on the ways in which variations in man's natural environment affect social development. Benton thus ascribes such an explanation to Marx when, as we saw earlier, he writes that the geographical variations of societies, as well as their historical transformations, are to be understood in terms of the ways in which they interact with nature.[94] Similarly, Plekhanov incorporates geographical explanation within historical materialism when he writes that 'the development of the forces of production (which for its own part, in the last resort, determines the development of all social relations) itself primarily depends upon the peculiarities of the geographical environment'.[95] Blackburn, however, finds such explanations missing from Marxism. Its error, he claims, is to treat geography as a constant, as 'background noise', when in fact it is a variable 'whose impact can deflect the course of social development in decisive ways'.[96]

What is at issue here, it should be noted, is not the fact that human societies depend upon nature for their survival (though Blackburn's emphasis on the destructive aspect of nature is intended to suggest that humans are rather more vulnerable in this respect than is recognised by many, including Marx), but rather the extent to which they are shaped by geographical variations in their natural environments. Whether Marx underestimated this is hard to assess, but (claims about 'background noise' notwithstanding) he certainly acknowledges that such influences exist and can be significant. If the productive process depends on natural conditions in the ways that Marx's conception of the labour process implies, then the character of that process will be affected by variations in those conditions. Productivity, and in a capitalist society profitability, will be affected by the availability of raw materials, existence of appropriate 'contextual

---

[94] See text to note 11 above.   [95] Plekhanov 1937, p. 36. See also p. 32.
[96] Blackburn 1990, pp. 14–15.

conditions' and so on.[97] Such factors will affect not only the geographical distribution of particular productive activities but also the propensity of societies in different regions to undergo the technological development that for Marx is the condition for social progress. Marx notes, on the one hand, that difficult physical conditions can inhibit development, as in the case of 'Asiatic' societies where the need to maintain extensive irrigation systems led to a rigid division of labour and lack of private ownership which in turn were responsible for social stagnation, and on the other hand, that the incentive to development may be lost in circumstances in which needs may too easily be satisfied:

> Where nature is too prodigal with her gifts, she 'keeps him in hand, like a child in leading-strings'. The mother country of capital is not the tropical region, with its luxuriant vegetation, but the temperate zone. It is not the absolute fertility of the soil but its degree of differentiation, the variety of its natural products, which forms the natural basis for the division of labour, and which, by changes in the natural surroundings, spurs man on to the multiplication of his needs, his capacities, and the instruments and modes of his labour.[98]

A possible objection is that Marx's references to the influence of geography relate primarily to the early stages of historical development, and that he is less willing to admit such an influence at a later stage. However, Marx would surely be right to hold that geographical variations become *less* significant as society develops,[99] and, as his reference to their effects upon capitalist profitability indicates, he continues to allow them a significant influence in contemporary societies.

4.7 Narrow and broad historical materialism

In this chapter we have discussed a number of passages in which Marx highlights the significance for human societies of their natural environ-

---

[97] As Marx observes: '[i]f we assume capitalist production, then, with all other circumstances remaining the same, and the length of the working day a given factor, the quantity of surplus labour will vary according to the natural conditions with which labour is carried on, in particular the fertility of the soil' (*Capital*, vol. I, p. 648).

[98] *Ibid.*, p. 649. For Marx and Engels's account of 'Asiatic' societies, see 'The British Rule in India'.

[99] The reasons for this include the following. (1) More developed technology and infrastructure is better able to mitigate the effects of unfavourable geographical conditions – hence, for example, the differing impacts of natural disasters in more and less developed societies. (2) The development of international commerce means that a society's development is no longer dependent solely on its local resources and indigenously developed technologies. (3) International commerce and, more dramatically, colonialism, result in the disruption of stagnating social structures in regions where the conditions for spontaneous development are absent, and in their being sucked into the general pattern of capitalist development. (See, for example, Marx's judgement on the revolutionary effects of British colonialism in 'The British Rule in India', pp. 306–7.)

ments, both as conditions for their existence and as influences on their development. Blackburn's critique of Marxist 'productivism', discussed in the last section, suggests two responses to such passages. Firstly, and most explicitly, he suggests that Marx does not go far enough in recognising either the extent to which nature impacts (particularly in destructive ways) upon human societies, or the geographical variability of its impact. These charges were considered in the previous section. Secondly, however, Blackburn's characterisation of historical materialism as a 'productivist' theory[100] suggests that, for Blackburn, Marx's affirmation of the dependence of humans upon nature, and of the effects that geographical variations have upon human societies, while welcome in itself, falls outside his core theory of historical materialism. The general view here being ascribed to Blackburn is that historical materialism is a theory of economic or technological determinism, or at any rate a theory about forces and relations of production, base and superstructure, and not in a significant way a theory about the relation between nature and society.

One reason for hesitating to separate Marx's assertions of human dependence on nature from historical materialism in this way is his description – noted above – of the relation between humans and non-human nature, and the physical facts underpinning that relation, as 'natural bases' for the writing of history. This matter, however, is worthy of further investigation, since it has a bearing on the weight that should be attached to Marx's assertions of human dependence upon nature. Writers who interpret historical materialism in the narrow sense described above are not thereby committed to any particular view about the representativeness of Marx's quoted statements on nature and humanity. However, the exclusion of these statements from historical materialism means that they offer no reason to think that this theory, or the other parts of Marx's thought for which it serves as a 'guiding thread', will be able to account for the development of ecological problems. It also leaves open the possibility that Marx's stated opinions on the relation between nature and humanity may be contradicted by his theory of history. In particular, historical materialism, thus construed, appears vulnerable to the charge that its concept of the growth of the productive forces ignores limits on that growth resulting from humans' dependence upon nature. This charge will be examined in the next chapter. In what follows I will consider some arguments about the proper scope of 'historical materialism', and about the relation between that theory construed in the narrow sense described

---

[100] This may be contrasted with Benton's view that Marx's 'productivist' bias resides in his political economy, and that historical materialism itself is a 'naturalistic' and 'materialist' theory.

and Marx's more broadly materialist commitment to the principle of ecological dependence.

'Narrow' interpretations of historical materialism receive their strongest textual support from Marx's 1859 Preface, described by Cohen as 'the canonical text' for the interpretation of historical materialism. 'In attending to the Preface,' he writes, 'we are not looking at just one text among many, but at that text which gives the clearest statement of the theory of historical materialism.'[101] In the Preface, Marx says:

> In the social production of their life, men enter into definite relations that are indispensable and independent of their will, relations of production which correspond to a definite stage of development of their material productive forces. The sum total of these relations of production constitutes the economic structure of society, the real foundation, on which rises a legal and political superstructure and to which correspond definite forms of social consciousness.[102]

Cohen quotes this passage to illustrate the content of Marx's historical materialism, but omits the final clause referring to social consciousness.[103] This leaves three 'ensembles' that Marx refers to: productive forces, relations of production and superstructure. The definition of these ensembles and explication of the explanatory relations among them, for Cohen, represent the content of historical materialism.[104]

Although Cohen is by no means alone in treating the Preface as canonical,[105] there are reasons to be cautious about according it this degree of exegetical primacy. Firstly, there are reasons relating to the status of the text and the circumstances of its publication: it is after all 'only' a preface, in which Marx sums up, very briefly and schematically, the results of extensive previous work, and which according to one commentator he 'gladly allowed to go out of print, as superseded by later writings'.[106] It has also been suggested that Marx omitted aspects of his previous work – in particular the notion of class struggle – from this summary in order to get it past the Prussian censor.[107] Be this as it may, there is surely much that was omitted simply because of lack of space. Secondly, the text itself does not unequivocally define historical materialism in the way that Cohen's use of it suggests: Marx introduces the quoted passage not as a summary of his-

---

[101] Cohen 1988, p. 3.   [102] 1859 Preface, p. 181.
[103] Cohen 1978, p. 28; Cohen 1988, p. 4.
[104] See also Shaw 1978, p. 59. Such interpretations of historical materialism may be contrasted with Keith Graham's (1992, p. 10) contention that historical materialism is not 'just one thing', but 'a welter of different thoughts and ideas, not all of which have the same status'.
[105] Cf. Shaw 1978, pp. 8, 55; Shaw 1991, p. 235. Plamenatz (1963, pp. 18–20) describes the Preface as 'the classic formulation of historical materialism'.
[106] Miller 1984, p. 175.   [107] See Prinz 1969. Also Moore 1975, p. 174.

torical materialism (or any equivalent phrase), but as the *general result* of his *investigation of political economy*. Neither the contextual nor the textual point is conclusive in itself. Regarding the former, it has been pointed out that Marx himself was happy to quote from the Preface in his later works including *Capital*.[108] Regarding the latter, it may be noted that Marx elsewhere describes the quoted passage from the Preface as a discussion of 'the materialist basis of my method',[109] and that Engels, in his review of *A Contribution to the Critique of Political Economy*, describes the doctrine outlined in its Preface as the 'materialist conception of history' and the 'essential foundation' of Marx's political economy.[110] The point is that Marx appears not to have been as definite about the compartmentalisation of his corpus into discrete components as Cohen's (and others') definition of 'historical materialism' would suggest. It may therefore be useful, rather than attempting to stipulate a definition, simply to distinguish between historical materialism narrowly defined, as for example by Cohen, and a broader definition which would include Marx's commitment to the principle of ecological dependence, as documented in section 4.2. The question to be answered in this section can then be formulated in terms of the relation between historical materialism narrowly construed, and those elements of the broader theory which concern human dependence upon nature.

Cohen argues against what he sees as the misconception by Marx, and many of his followers, that an aspect of humans' dependence on nature, namely their need to eat, shelter and clothe themselves, is sufficient to demonstrate the truth of the narrow theory of historical materialism.[111] His critique focuses on the argument presented by Engels in his obituary address at Marx's funeral. Engels said:

Marx discovered the law of development of human history: the simple fact, hitherto concealed by an overgrowth of ideology, that mankind must first of all eat, drink, have shelter and clothing, before it can pursue politics, science, art, religion, etc.; that therefore the production of the immediate material means of subsistence and consequently the degree of economic development attained by a given people or during a given epoch form the foundation upon which the state institutions, the legal conceptions, art, and even the ideas on religion of the people concerned have been evolved, and in the light of which they must, therefore, be explained, instead of *vice versa*, as had hitherto been the case.[112]

According to Cohen, Engels is here guilty of equivocation between two senses in which the production of food (etc.) might be described as the

---

[108] Shaw 1991, p. 235; Graham 1992, p. 80n. Cf. *Capital*, vol. I, p. 175n.
[109] *Capital*, vol. I (Postface to the Second Edition), p. 100.
[110] 'Karl Marx, A Contribution to the Critique of Political Economy', p. 469.
[111] Cohen 1988, pp. 124–31.   [112] 'Speech at the Graveside of Karl Marx', p. 429.

'foundation' of politics, law, etc. In the first sense, production of food is a *necessary condition* for engagement in politics; in the second sense the former is understood to *explain* the latter. In the first sense, Cohen argues, the statement that production of food is the foundation of politics follows (indeed, follows trivially) from the necessity of eating; in the second sense, however, it does not. Cohen therefore concludes that the narrow theory of historical materialism, which claims to explain state institutions, law, etc. by reference to the *way in which* we produce our needs, cannot be deduced from the mere fact that we are material beings who have needs for food, clothing and shelter.

Cohen is clearly right that Engels's statements, as presented, do not amount to a deductively valid argument. But is this what Engels intended? Keith Graham suggests that he 'is more plausibly interpreted as putting forward a hypothesis (with more certainty than is warranted) together with the considerations prompting it', or in other words as giving *'prima facie* grounds for thinking that material production may have explanatory primacy'. The necessity of producing food (etc.) counts as such a ground because the amount of time and effort that we must put into it makes it so central to our lives.[113] A similar view is taken by Andrew Collier, who argues that while Marx, Engels and Lenin tended to exaggerate the degree of support that narrow historical materialism may obtain from biological facts such as the need to eat, there is nevertheless a non-deductive connection to be made.[114] Cohen's argument is directed solely at those who would assert a relation of logical entailment from humans' possession of material needs to narrow historical materialism. He would not necessarily reject the suggestion that the former are, in some weaker sense, 'grounds' or 'foundations' for the latter, and indeed there are elements in his account that would suggest some such relation.

First, Cohen leaves open the question of whether Engels's argument could be made deductively valid by the addition of further premises.[115] If this is the case, and the required additional premises are ones which we have reason to think true, then it would follow that the need to produce is a reason (or at least part of a reason) to think narrow historical materialism true. Second, the theory of narrow historical materialism, in Cohen's interpretation and for Marx himself, would not apply to a world in which humans were not dependent upon nature and did not need to produce in order to satisfy their material needs.[116] Human dependence upon nature is,

[113] Graham 1992, pp. 14, 15.   [114] Collier 1979, esp. pp. 42, 53–4.
[115] Cohen 1988, p. 128.   [116] Cf. Collier 1979, p. 44.

in other words, a necessary though insufficient condition for the truth of narrow historical materialism. This is plausibly what Marx has in mind when he speaks of physical facts about human and non-human nature as the 'natural bases' of history, and it is certainly implicit in his view that historical development will not occur 'where nature is too prodigal with her gifts'.[117] In Cohen's own exposition of historical materialism, the way in which relations of production and in turn superstructure are explained by the level of development of the productive forces relies (for reasons that will be considered in section 5.5) upon the thesis that '[t]he productive forces tend to develop throughout history', and this in turn is explained by, amongst other things, the fact that humans exist in a situation of scarcity in which it is rational for them to seek new techniques to improve their situation.[118] Scarcity, in this context, amounts to the dependence of humans upon a less than 'prodigal' nature. It follows, therefore, that even if Marx's assertions of human dependence upon nature are excluded from the content of a narrowly defined historical materialism, the fact of such dependence remains indispensable to the applicability of that theory and therefore to Marx's broader work for which narrow historical materialism formed the 'guiding thread'.

4.8 Conclusion

In section 4.2 we saw that Marx repeatedly affirms something very like what I have called the principle of ecological dependence. He does so in contexts which leave no doubt that Marx himself thought this general materialist principle to have profound methodological importance for his narrower theory of history. The subsequent sections have examined various challenges to this assumption. These challenges, I have argued, fail to show that Marx abandoned or contradicted his commitment to the principle, but they do indicate further questions that need to be addressed.

The claim that Marx abandoned the principle of ecological dependence in his later works, I argued, is contradicted by assertions of the principle that occur throughout his mature writings. I suggested that the reason such an abandonment is alleged, is that another important feature of the later works – the notion of the development of the productive forces, central to narrow historical materialism – is thought to be incompatible with the principle. I have argued, however, that the transformation of nature by humans is not necessarily incompatible with the principle of ecological dependence, and that both are necessary components of an

---

[117] See text to note 98 above.   [118] Cohen 1978, pp. 134, 152.

adequate conceptual framework for understanding ecological problems. Further investigation is therefore required in order to establish whether Marx's particular account of human transformation of nature is in conflict with his recognition of human dependence upon nature. This investigation will be undertaken in the following chapter, where I will consider the interpretation and ecological significance of Marx's commitment to the development of the productive forces.

We have also seen that the concept of human need is central to an understanding of the connection between Marx's broader materialist understanding of the human–nature relation and his narrower theory of history. The dependence of humans upon nature is expressed in terms of their possession of material needs, and the satisfaction of those needs is the underlying purpose of the labour process and the motivation for developing the productive forces.[119] This is noted by Collier, who describes 'the presence of needs, with definite ranges of variation in means to satisfaction' as both 'the point of contact between the human sciences and their biological foundation', and, within the human sciences, as 'the foundation of the determination in the last instance by ... the "material" element', and by Graham, who identifies the possession of material needs that are universal and recurrent, and whose satisfaction consumes much effort and time, as the factor that links Marx's 'basic' materialism with the explanatory primacy of material production.[120] The concept of human needs also figures strongly in green writings, but here, in accordance with the aim of reducing material consumption, the tendency is to minimise the extent of human needs, and to distinguish needs from mere wants. Communism, as well as capitalism, is criticised for its alleged commitment to a continuous expansion of its population's needs, and an expansion of production to match them.[121] This issue will be addressed in the final chapter where I will investigate Marx's conception of human needs and its ecological implications.

---

[119] It is the *underlying* purpose and motivation, because the immediate purpose, in the case of commodity production, is the production of profit. In other words it is Marx's account of the valorisation process rather than the labour process which captures the immediate motivations of those who control production.

[120] Collier 1979, pp. 44–5; Graham 1992, pp. 14–15.   [121] Dobson 1990, pp. 18, 29.

5 Development of the productive forces

We have seen that Marx is committed to the view, which forms a corner-stone of environmental thought, that humans are dependent upon their natural environment. This view, I argued in the last chapter, is not contra-dicted by the idea that it is characteristic of humans to transform their envi-ronment. Nor, I argued in chapter 2, is a striving for economic or productive growth *per se*, in conflict with a recognition of the finitude of the natural resources upon which the species depends. It is clear, however, that of the different ways in which humans may transform nature, and expand their transformative powers, only some will be compatible with the avoidance or minimisation of ecological problems. It is therefore not sufficient to consider transformation of nature in the abstract; what is required is an examination of the particular account of the transformation of nature given by Marx, at the centre of which is his notion of the devel-opment of the productive forces. My aim in this chapter is therefore to con-sider how this notion should be understood and the role that it plays in Marx's theory, and to address the contention of many of Marx's green critics that his use of the notion leads inevitably to the exacerbation of eco-logical problems. More speculatively, I will suggest reasons why the notion may be a productive one for investigating ecological problems.

The first thing to notice is that the development of the productive forces plays both an explanatory and a normative role for Marx. It plays an explanatory role in that the development of the productive forces is what for Marx explains the transition from one social structure to another. Insofar as the development referred to in this explanation is intended merely as a description of what actually happens, there is no cause to crit-icise Marx for any tendency it might have to produce or exacerbate ecolog-ical problems; indeed Marx's account of the development of the productive forces, and the reasons for its occurrence, may be construed as an explanation of contemporary ecological problems. However, as we shall see in more detail later, the development of the productive forces is not, for Marx, a process that occurs independently of human actions, but a product of actions and structures chosen by human beings. This raises the possibility of promoting, restraining, or possibly redirecting that process. Marx's own positive evaluation of the development of the productive forces, even when it is achieved at great human cost, can be seen in the praise that he gives capitalism in the *Communist Manifesto* for having pushed forward that development. Three reasons may be discerned for this positive evaluation. The first is that development of the productive forces makes it possible for people to produce more of what they need with less effort and in less time, even if under capitalism this potential is only partly realised. The second, to be considered further in chapter 6, is that the exercise and expansion of humans' creative powers, exemplified (albeit in a distorted manner under capitalism) by the development of the productive forces, is for Marx an important part of human nature and a component of human flourishing. And third, a high level of development of the productive forces is for Marx a necessary condition for the establishment of communism, which he sees as desirable not only because it allows the productive forces to develop further than is possible under capitalism but also because it is a more just and less alienating society, more suited to the creative and social aspects of human nature. Thus, to the extent that Marx views development of the productive forces normatively, as a necessary condition for the realisation of his projects, he is vulnerable to the claim that those projects are tainted – and possibly undermined – by ecological problems associated with that development. It is therefore necessary to consider both the kinds of productive development to which Marx is normatively committed and the ecological consequences of that development.

5.1 Development of the productive forces and development of technology

To answer the question of why development of the productive forces might be expected to be ecologically problematic, we need to know what it is whose development we are considering. Marx himself never explicitly defines the productive forces, but the following plausible and widely held view may be derived from his account of the labour process, discussed in the last chapter. The productive forces consist of *labour power*, and the *means of production* that labour power utilises in order to make its products. Labour power is comprised of an agent's strength, skill, knowledge, inventiveness, and so on, while the means of production consist of *instruments* and *objects* of labour.[1] Since ecological problems are problems arising out of humans' dealings with nature, it is the natural components of the means of production that concern us here, and, as we saw in the last chapter, Marx's account of the labour process makes it very clear that both the instruments and objects of labour originate from nature and have a persisting natural component. Objects of labour (or *raw materials* as I will henceforth call them)[2] are either given directly by nature or are natural objects modified by previous labour processes. Instruments of labour include: natural objects such as stones used as tools in primitive labour processes; tools and machines manufactured out of natural materials; and even the earth itself which serves as an instrument of labour in agriculture. And, as noted in chapter 4, it is not only things which directly 'conduct' the worker's activity onto its object which Marx recognises as instruments of labour; he also defines as instruments of labour in a wider sense 'all the objective conditions necessary for carrying on the labour process',[3] a category which he intends to cover such things as workshops, canals and roads, but which will also without modification include the natural systems, physical, biological and climatic, upon which production depends.

These natural components of the production process (shown schematically in figure 5.1) indicate two aspects of the process that make it liable to produce ecological problems:

(i)   its dependence on *naturally given raw materials*, and
(ii) its dependence on *naturally given instruments of production*.

In addition, it was noted in the last chapter that of the materials used in the productive process, only a part ends up in the product, and only some of these materials' properties are understood and exploited by the producers. The production process is therefore also liable to ecological problems in virtue of:

---

[1] See Cohen 1978, ch. 2, especially pp. 32 and 55. Cohen argues that what links the items falling under these two categories is they are all, in some sense, '*used* by producing agents to make products' (p. 32). Someone, he adds, must *intend* that the item contribute to production in this way, but it is not necessary that that someone be the direct producer. Cohen departs from Marx in adding *spaces*, alongside instruments and objects of labour.

[2] As noted in the previous chapter, this usage departs from Marx's. Whereas Marx reserves the term 'raw materials' for those objects of production that have passed through a previous labour process, I am using it in its everyday English sense – equivalent to Marx's 'objects of labour' – to cover both natural and previously processed materials.

[3] *Capital*, vol. I, pp. 285–6.

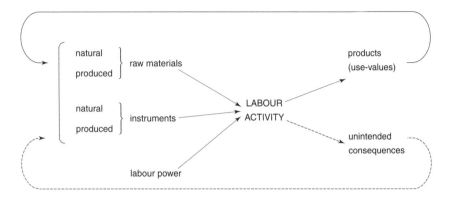

Solid arrows represent connections noted explicitly in Marx's account of the labour process; broken arrows represent connections noted elsewhere.

Fig. 5.1. Ecological impact of the labour process.

(iii) its production of pollution and other environmentally detrimental *unintended consequences*.[4]

If the development of the productive forces entailed an expansion of each of these ecologically problematic elements – consuming more from nature and having greater unintended impact upon it – then the claim that such a development must inevitably add to our ecological problems would appear to be confirmed, and Marx's green critics vindicated. But the premise of this argument warrants further scrutiny. It has yet to be demonstrated that expansion of these ecologically problematic elements is a necessary consequence of the development of the productive forces, and in what follows I will give reasons for doubting that such a tight connection can be made.

How, then, is 'development of the productive forces' to be understood? Marx's green critics often perceive historical materialism as a form of technological determinism, and this suggests that they equate development of the productive forces (the 'determining factor' in narrow historical materialism) with the development of technology. This characterisation of historical materialism is disputed by many Marxists; however, leaving aside

---

[4] I am using 'unintended' in a broad sense to include consequences of an action which may be foreseen but are not part of the purpose of the activity (cf. note 2, ch. 4 above). The effects of the production process upon the environment may be judged undesirable either because of their direct effects upon humans (e.g. in terms of health or aesthetic or recreational enjoyment of nature) or other creatures, or because they form a feedback loop, reacting upon the natural preconditions (objects or instruments) of the productive process and undermining its sustainability.

the question of determinism for now, I will argue that there is a close connection to be drawn between development of the productive forces and the development of technology, provided that the latter is understood in a suitably broad sense. I will, however, argue against the claim, also put forward by some of Marx's green critics, that technological development leads inevitably to a worsening of ecological problems.

Some commentators equate technology with tools and machinery, that is, with instruments of production.[5] Understood in this way, 'development of technology' appears to be a narrower concept than Marx's 'development of the productive forces', since the productive forces include labour power as well as instruments. But the development of labour power consists primarily[6] in the development of skills and knowledge, and this corresponds to what several commentators have noted is the original meaning of 'technology': 'knowledge about technique' or 'knowledge of the industrial arts'.[7] (Skills, I am assuming, will be included within this characterisation of technology, under the category of practical knowledge, or 'knowing *how*'.) In fact, any actual development of technology must involve *both* the material and the human elements, since each depends upon the other: the development and application of tools and machines are impossible without the knowledge to invent them and the skills to use them,[8] and conversely the development of labour power is capable only of limited advance until it avails itself of new instruments of production. The development of technology may therefore be taken to include the development both of the instruments of production, and of labour power.

This broadened account of technology still appears to leave one element of the productive forces – raw materials – unaccounted for.[9] I think, however, that insofar as Marx's phrase 'development of the productive forces' can meaningfully be applied to raw materials, it must refer to such

---

[5] See, for example, Grundmann 1991b, p. 107.

[6] 'Primarily', since the development of labour power might also include development of the physical strength of workers.

[7] Freeman 1977, p. 225; Layton 1977, p. 199; cf. Mitcham 1979.

[8] Cohen (1978, p. 42) argues that knowledge, rather than physical instruments, is the more important part of technology, since instruments can be rebuilt given sufficient knowledge, but without the requisite knowledge the instruments become useless. This thought lies behind Marx and Engels's comment, in *The German Ideology*, p. 72, on how the communications forged by international commerce assure the permanence of productive forces, which, while they remained local, were vulnerable to complete destruction. Marx also emphasises the priority of knowledge over material objects in *Grundrisse*, p. 706 where he describes machinery, etc. as 'the power of knowledge, objectified'.

[9] Naturally given instruments of production are also excluded, insofar as they remain undeveloped, but as soon as they undergo development and become artefacts they may be included within the category of technology.

events as the discovery of new resources, of techniques to increase the viability of marginal deposits, and of new uses to which raw materials can be put, developments which are largely made possible by innovations in the instruments and techniques (i.e. the technology) of surveying, extraction and manufacture. It is true that the phrase 'development of resources' is sometimes used to refer to the process of extracting minerals from the ground, but a routine part of the productive process like this cannot be an example of development of the productive forces in the Marxian sense, since this is understood as some sort of historical advance, a *change* to the routine of the productive process. It would also be counter-intuitive to construe 'development of the productive forces', as applied to raw materials, as referring simply to an increase in the rate at which raw materials are consumed (although an ecologically unaware Marxism might celebrate as a development of the productive forces the technology that makes faster extraction and consumption possible). Given, then, that 'development of the productive forces', as applied to raw materials, must refer to developments of the instruments and techniques (i.e. the technology) with which the materials are discovered, extracted and used, it follows that such developments fall squarely within the account of technological development given above.

What these considerations show is not that 'development of the productive forces' *means* 'development of technology'. This cannot be the case, since technology (or, more specifically, productive technology) is only a part of the productive forces, which additionally include the physical capacities of workers, raw material and naturally given instruments of production, and since it is possible for productive technology to develop without an overall expansion of the productive forces, or for the productive forces to decline while technology remains stationary or even advances, if there is a decline in the non-technological elements of the productive forces such as the stocks of raw materials. Rather, what has been shown is that, as a matter of fact, it is only by developing the technological elements of the productive forces that humans are able significantly to expand their productive capacities. It follows from this that if the development of technology gives rise inevitably to ecological problems, as some critics contend, then those problems will be similarly associated with the development of the productive forces. In the following section I will therefore consider some of the grounds that have been put forward for the view that technological development (and hence development of the productive forces) must inevitably be ecologically damaging.

5.2 Technological development and ecological problems: an inevitable correlation?

New technology may, and quite often does, waste more raw material, burn more fuel, and produce more pollution than the technology it replaces.[10] The question, however, is not whether such consequences are possible, but whether they are an inevitable accompaniment to all forms of technological development, or whether, on the other hand, there are forms of technological development, satisfying Marx's criteria for the development of the productive forces, which can avoid such consequences.

One argument concerning the ecologically problematic nature of technological development is given by Reiner Grundmann, who highlights particular features of modern technology as evidence of its propensity to cause ecological problems. Grundmann's first point is that the possible range of ecological damage is greatly increased as the 'scope' of technology grows: 'Mankind in its early stages, with primitive technology, could not affect its environment in the same way as mankind can today: the axe and fire could not, even under conditions of most careless use, cause dangers which were in the least comparable to present dangers which arise out of the use of nuclear or chemical technology.'[11] Increased scope of technology, as invoked here, would appear to be central to any notion of technological development, but in itself this only creates the *possibility* of increased ecological problems. It may be argued that, though new technologies increase the potential for ecological damage, they may also provide more effective means of controlling technology which will reduce the probability of that potential ecological damage actually occurring, so that overall the likely level of environmental impact is lowered rather than raised. An example of this type of argument is the claim made by the nuclear power industry that nuclear power generation is 'cleaner' in operation (i.e. emits less atmospheric pollution) and statistically causes fewer deaths than coal-fired power generation.[12] In response to this, it may be argued that *whatever* precautions are taken, and whatever safety devices are in operation, the probability of (for example) nuclear or chemical

---

[10] The harmful effects of many twentieth-century technological developments are chronicled in Commoner 1971.   [11] Grundmann 1991b, p. 29.

[12] The suggestion that technological developments may improve control and hence reduce the incidence of environmental damage and the probability of accidents may be thought particularly pertinent to Marxism, for which the question arises not only of improved technical means for monitoring and controlling environmental impact, but also of technological development ushering in new social forms that may be better able to control its effects. This of course is disputable, and although I will not address this question in detail, I will have more to say about it later in this chapter and in the conclusion.

pollution will still be greater than it was before those technologies were invented, i.e. greater than zero, and moreover, any non-zero probability represents a statistical inevitability in the long term. This may be a reason for rejecting technologies such as nuclear power, where the consequences of an accident are extremely grave, however low the probability of an accident,[13] but in general this response misses the point that powerful and potentially dangerous technologies already exist, and though we could perhaps do without any individual one of them, we could not – given current population levels, the dependence of our production and infrastructure upon electrical power, and so on – do without them all. The situation, therefore, is that there is already a risk (or worse) of serious environmental damage – for example from nuclear accidents or the global warming resulting from the routine emissions of coal-fired power stations – and in this context the question to be addressed is not whether new technologies carry dangers (they do!), but whether those dangers are better or worse than the dangers carried by the existing alternatives.[14]

It is here that Grundmann's second point comes in. He argues that the complexity and 'tight coupling' of modern technologies respectively reduce the transparency of their operation and allow small failures to produce large effects, with the consequence that environmental damage in the course of normal operation becomes more likely, and accidents like Three Mile Island, Chernobyl or Bhopal become inevitable, irrespective of any safety devices that might be employed.[15] It may be that Grundmann overstates this case; in particular it is not clear that he gives sufficient weight to the countervailing potential of sophisticated control and safety systems, which may be more able than humans or more primitive control systems to deal with complex technology, and in the light of this it is not clear that the inevitability referred to is anything more than that of any event of finite probability, given enough time. But even if we accept his contention that ecological problems inevitably result from complex and tightly coupled technology, it does not follow that such consequences necessarily result from the development of technology. This is because increased complexity and tightness of coupling are contingent features, not essential to the development of technology. The possibility of alternative paths of technological development is in fact recognised by Grundmann, when he addresses the question of whether, or to what

---

[13] Cf. Grundmann 1991b, p. 35.

[14] Also, there are environmental dangers that do not result from humans' use of technology, but from natural disasters such as floods, earthquakes or meteor collisions. Possession of powerful technologies may help us to avert these dangers, which should therefore be offset against the dangers of those technologies.   [15] Grundmann 1991b, p. 33.

extent, it is possible for societies to direct their technological development and to bring into existence technologies which are less likely to damage the environment.[16]

An argument which does purport to establish a necessary connection between technological development and ecological problems, and is representative of much green thought, is advanced by Val Routley. She argues that the vision of an 'automated paradise' offered to us by Marx 'must be highly energy-intensive and thus given any foreseeable, realistic energy scenario, environmentally damaging'.[17] Routley's assumption of a correlation between automation and resource consumption appears to be shared by G. A. Cohen, who argues that the 'resources crisis' could be thought to make Marx's post-capitalist project look naive, '[f]or if natural resources are to be used more sparingly, recourse to them must to some extent be replaced by continued reliance on human labour power, and it might then seem that the promise of increased leisure time cannot be fulfilled'.[18] Cohen counters this as a criticism of Marx by questioning the assumption that, for Marx, leisure time is necessarily unproductive. Cohen is right to question this assumption, as we will see in chapter 6, but both he and Routley are wrong to assume that an increase in automation must always result in an increase in energy consumption.

Automation of production involves two elements: the replacement by machines of mental labour and of manual labour.[19] The latter does indeed require the substitution of natural sources of energy for the energy previously supplied by human labour; thus the transition from handicraft production to machine industry implies an increase in the requirement for

---

[16] *Ibid.*, pp. 140–1; cf. p. 182. This is central to Grundmann's account. He thinks that the fact that there are some forms of technology that will be ecologically problematic under *any* social form undermines the importance of relations of production in explaining ecological problems, and calls for a shift of political focus to the question of whether and how technological development can be directed. We may wonder, however, whether these questions really are as separate as Grundmann supposes, or whether social form affects the direction (or the possibility for controlling the direction) of technological change. I will comment further on Grundmann's account and his reasons for discounting the effects of relations of production in my conclusion.   [17] Routley 1981, p. 242.

[18] Cohen 1978, p. 323.

[19] Automation is defined, in the *Concise Oxford Dictionary*, as 'Automatic control of manufacture of product through successive stages; use of automatic equipment to save mental and manual labour.' The reference to *controlling* the process, in the first part of this definition, suggests that the replacement of mental labour is more central to the concept than the replacement of manual labour, a view which seems to me to accord with the usual use of the term, and to be required if we are to make sense of the distinction between 'machinery' and 'automatic machinery'. However, since real cases always involve the replacement of both mental and manual labour, in varying proportions, and since both elements are clearly present when the term is used to express Marx's understanding of the development of technology, both elements must be considered.

natural energy resources. But if the starting-point is today's highly mechanised production, which already is heavily reliant on such resources, the picture is different. In this context, automation must, to a large extent, mean the substitution of machines for the predominantly mental labour which humans expend in controlling or supervising machine production. In this capacity, human labour contributes little to the overall energy requirements of the productive process, and it is quite possible that an automatic system will operate in a more energy-efficient way than it would under direct human control, more than offsetting the relatively small amount of extra energy required to replace the supervisory labour. To take a familiar example, the energy required to perform a gear change on a car is small compared to that required for propulsion; so an automatic gearbox that selected the most energy-efficient gear more consistently than a human operator, could yield a reduction in fuel consumption greater than the small amount of extra fuel consumed by the mechanical action of changing gear.[20]

Routley's focus on energy use is typical of much ecological argument, but similar considerations apply to other resources. Just as the development of automation may lead to a net saving in energy consumption, so it may lead, by similarly improving efficiency, to savings in the use of other types of resource (i.e. raw materials), and to reduced emission of pollutants. This indicates that development of technology has a part to play in dealing with environmental problems; it does not, however, license the assumption that technological changes alone will be sufficient to resolve ecological problems, since there are theoretical limits to what can be achieved in the way of increased efficiency, and even where improvements are theoretically possible there is no guarantee that the technological means to achieve them will be discovered in time to avert ecological problems, if indeed at all.[21]

---

[20] My purpose here is not to suggest that automation of this kind can eliminate the ecological destruction associated with present levels of motor car use, but simply to illustrate the fact that the automation of technology envisaged by Marx need not be ecologically destructive in the way that Routley assumes, but may contribute (albeit in a limited way, as I indicate in the following paragraph) to a reduction in its ecological impact.

[21] This cautionary note constitutes the rational core of green objections to 'technological fixes' to ecological problems. Limits to the technological amelioration of ecological problems are determined, for example, by the quantity of materials and energy contained in the product and necessary for the transformation of the raw material into the product. These minima of inputs will also determine minima of waste products, constituted by that part of the fuel or raw material input that does not become incorporated in the product. Technology may, however, enable waste products to be converted into less harmful forms or to be used as resources in some other process.

Many greens do of course recognise that despite these reservations, technology has a contribution to make. A useful survey of green attitudes is contained in Spretnak and

The automation to which Routley (mistakenly) objects consists of an increase in the productivity of labour, i.e. an increase in the ratio of size of product to the amount of direct labour required to produce it, where this is achieved by means of technical innovation in such a way as to reduce the need for labour power.[22] For others, however, the idea that the development of technology necessarily increases its impact on the environment rests on the assumption that the purpose of increasing productivity is to permit an increase in the quantity of goods produced, by loosening the constraint previously imposed by the requirement for labour power.[23] Such a development would, other things being equal, lead to an increase in the quantity of resources consumed and an increase in the quantity of waste products. But as we have seen, other things are not always equal, because these consequences may to some degree be offset if the technology that increases labour productivity is also more ecologically efficient; it may therefore be possible to produce more with the same resources or even to produce more with less. Even so, given the limited scope and uncertainty of such efficiencies, development of technology aimed at an increase in production must be considered potentially problematic.

We have considered cases where technological development is aimed at increasing labour productivity, either by increasing the size of the product or by reducing the requirement for labour. However, it need not be aimed at increasing labour productivity at all. For even if the greens' proposal for a halt to the growth of production was enacted, and it was decided to forego further reductions in labour time, it would still be rational to introduce technological innovations designed to increase the efficiency with which resources are used and waste products disposed of, in order to reduce the ecological impact of existing levels of productive activity. Such innovations would appear to deserve the title of 'technological development' no less than innovations aimed at maximising labour productivity.

5.3 Criteria for technological development

What we have seen in the previous section is that technological development can take different forms and serve different purposes, with differing ecological consequences. But can this diverse range of objectives really be

Capra 1986, p. 88. However, the widespread use of the terms 'hard' and 'soft' to designate environmentally damaging and environmentally benign technologies does suggest a preference for traditional or 'intermediate' technologies over modern 'high' technology. There is certainly a place for the former but in many cases it will be the latter that is more environmentally efficient. See *ibid.*; also Irvine and Ponton 1988, p. 48.
[22] This formulation comes from Cohen 1978, p. 56.   [23] E.g. Porritt 1985, p. 44.

combined within a single concept of 'technological development'? Consider first the account proposed by Henryk Skolimowski:

> The growth of technology manifests itself precisely through its ability to produce more and more diversified objects with more and more interesting features, in a more and more efficient way.
>
> It is a peculiarity of technological progress that it provides the means (in addition to providing new objects) for producing 'better' objects of the same kind. By better many different characteristics may be intended, for example: (1) more durable, or (2) more reliable, or (3) more sensitive (if the object's sensitivity is its essential characteristic), or (4) faster in performing its function (if its function has to do with speed), or (5) a combination of the above. In addition to the just-mentioned five criteria, technological progress is achieved through shortening the time required for the production of a given object or through reducing the cost of production. Consequently, two further criteria are reduced expense or reduced time, or both, in producing an object of a given kind.[24]

In this passage, we can discern three main features which, in Skolimowski's view, constitute developments of technology:

(i)   production of new kinds of object;
(ii) improvement of existing kinds of object (of which (1) to (5) are examples); and
(iii) reduced costs (in time and money) of production.

However, Skolimowski's further elaboration of the nature of technological progress focuses just on (ii). Feature (iii) is relegated to a secondary role while (i) drops out of the picture altogether.

Despite his pluralistic account of what constitutes a 'better' object of a given kind, Skolimowski thinks that the various criteria – durability, reliability, sensitivity, etc. – can be brought under a common label, as the kinds of 'effectiveness' appropriate to different branches of technology. In surveying, he says, the decisive element which guides the map-maker in his choice of techniques is accuracy of measurement. In civil engineering it is durability of construction, and in mechanical engineering it is efficiency (in the narrow technical sense). Skolimowski thus defines technological progress as: 'pursuit of effectiveness in producing objects of a given kind'.[25]

Skolimowski's examples, however, obscure some of the complications that arise when we consider technological progress in its real context. Firstly, extrinsic considerations such as economics or ecology, rather than the qualities of the object produced, may be of prime importance; and sec-

---

[24] Skolimowski 1983, p. 44.   [25] *Ibid.*, p. 45.

ondly, the object itself may be judged according to a multiplicity of criteria, rather than a single measure of effectiveness characteristic of that branch of technology.

One way in which Skolimowski attempts to reconcile these complications with his account is to maintain that all other measures of technological progress are subordinate to the one characteristic measure. Thus, in assessing the most economic method of surveying, 'the silent assumption is that the accuracy remains the same, or at any rate, that the decrease in accuracy is negligible', and in mechanical engineering, 'either we attempt to increase the absolute efficiency and raise it as close as possible to 1, or we attempt to construct a "better" engine while keeping the same efficiency (better can mean: safer, cheaper, longer lasting, more resistant)'.[26]

Skolimowski also attempts to define away complexities that threaten to undermine his thesis. Activities whose goals deviate from the specified measure of effectiveness are eliminated from the branch of technology in question, and criteria that compete with the specified measure of effectiveness are classed as non-technological. Thus, he argues that architecture is only in part technology and differs from civil engineering in that aesthetics and utility are at least as important to it as durability,[27] and, more generally: 'The analysis of the structure of thinking in technology is hampered by the fact that nowadays the construction of bridges, highways, automobiles, or even domestic gadgets is inseparably linked with the consideration of beauty and comfort which are basically "nontechnical" categories.'[28]

Skolimowski's attempts to deal with the complexity of real-world technologies within his framework are not convincing. To justify his use of the term 'nontechnical', he notes that aesthetic satisfaction and comfort 'cannot be measured objectively for all epochs'. But while this may create difficulties (perhaps even insuperable ones) for the measurement of technological progress, it does not in itself render these criteria 'nontechnical' in any relevant sense. For many technologies, aesthetics and comfort are not dispensable characteristics but key design concerns.[29] Skolimowski's elimination of counterexamples by reference to the specified measure of effectiveness renders circular his claim to have derived the measure from the actual practice of the technology. Thus his identification of a primary

---

[26] *Ibid.*, pp. 47, 48.   [27] *Ibid.*, pp. 46–7, 48.   [28] *Ibid.*, p. 49.

[29] For example, comfort must be a prime concern in the design of suspension systems for passenger transport, and in the technology of clothing textiles, manufacturers have sought to produce synthetic fabrics which are waterproof and breathable (comfort) yet have the appearance of natural fibres (aesthetics).

measure of effectiveness for each particular technology, and his relegation of other factors to a secondary role, is at best a stipulation of what he takes to be the 'essence' of that branch of technology; a stipulation which limits what is to be counted as progress without giving due weight to the range of purposes which innovation in that branch of technology may serve.

A more promising account of technological development is offered by I. C. Jarvie. He rejects the view that the appropriate measure of technological development in each case is determined by the branch of technology with which we are dealing, or the essential nature of its product, stressing instead the social character of technological problems. For example, Jarvie argues that durability is not necessarily the aim of building: 'Americans *could* build fireproof, floodproof, typhoon-proof, and earthquake-proof houses, but they are discouraged by the expense.' For engineers building a bridge to move an armoured column across a river, *speed of construction* may be far more important than durability. In surveying, *'accuracy* is only a relative aim, relative, that is, to the problem the surveyor and his cartographer are set. Terrain maps may ignore land use, road maps and tenancy maps may ignore terrain, maps of London and New York subways ignore all but the most general features of what they are depicting'. And the designers of (1960s) American cars 'certainly are not much concerned with the efficiency of their product so much as the appearance, the smoothness of ride, and so on. American cars could be much more durable and much more efficient (in terms of cost per mile) were manufacturers prepared to raise the unit cost'.[30] Jarvie concludes that the overriding aim, and the appropriate criteria for measuring technological progress, depend upon the concrete problem that is posed to the technologist, a problem that is always posed in a social context: 'Whether the overriding concern is with accuracy, durability, efficiency, or what, is always dictated by the socially set problem and not the technological field.'[31]

Jarvie is surely right in his insistence that the aims pursued by means of technological innovation are various (even within a single branch of technology), and that they are determined in large part by the social context in which that innovation takes place. The suggestion that the development of technology be measured according to its achievement of those aims also appears promising, not least in raising the possibility of including the costs of production, broadly conceived to include ecological costs, in the assessment of technological progress. Whether this possibility is realised will depend upon whether ecological matters fall within the 'socially set problem'. This, however, is a problematic notion.

[30] Jarvie 1983, p. 52.   [31] *Ibid.*

The difficulty is that, in the absence of a consensus among members of society, it is unclear how 'the socially set problem' should be defined. Different members of society, with different interests and preferences, are likely to judge technological innovations according to different criteria, or at least to attach different weights to the various criteria. For some the most significant factor will be the qualities and durability of the product while for others it will be the monetary costs of production, the safety and well-being of the workers who produce it, or the level of ecological impact. Thus, at a normative level, what counts as development is essentially contestable. It may however be useful for explanatory purposes to stipulate a non-normative definition – for example by defining Jarvie's 'socially set problem' in terms of the aims of those who actually control technological innovation. What is clear, however, is that even on a non-normative interpretation, 'technological development' may be defined in various ways, corresponding to the various aims that it may serve. The choice of such a definition must therefore be justified in terms of the explanatory task for which the definition is required. Our concern is not with technological development as such, but with the kinds of technological development that are implied by Marx's concept of the development of the productive forces. In order to answer that question we must therefore move our focus away from technological development itself and onto the latter concept and the role that it plays in Marx's theory of history.

5.4 Productive development in Marx: the Revolutionary Effect

We have seen that there are various factors which may in different contexts be regarded as criteria of technological development, and that the types of technological development defined by these criteria may differ in their ecological consequences. The question to be addressed here is how these various forms of technological development relate to Marx's concept of the development of the productive forces. Marx himself does not explicitly address the possibility of different paths of technological development. In his theory of historical materialism he writes as if the notion of development of the productive forces were unproblematic, and in his political economy of capitalism (the only social system whose structure he elaborates in detail) he takes this development to be equivalent to an increase in labour productivity, driven by the profit motive. However, what is important is not what Marx *thought* were the characteristics of the technological development that he believed to at the root of social progress, since we know already that his explicit pronouncements do not constitute an adequate response to contemporary ecological problems. What is important from an ecological perspective is to establish which forms of technological development Marx may be committed to (or barred from) endorsing by the requirements of his theory. To do this we need to consider the role that the development of the productive forces plays within that theory.

Marx's best-known discussion of the role of the productive forces is in the 1859 Preface, where he writes that the 'relations of production correspond to a definite stage of development of [the] material productive forces', a doctrine that he had earlier expressed in the aphorism '[t]he hand-mill gives you society with the feudal lord; the steam mill, society with the industrial capitalist'.[32] The productive forces thus serve in some sense to explain the prevailing relations of production (which in turn explain the legal and political superstructure and forms of social consciousness). Marx's primary interest, however, is not in explaining individual social forms, but in the *transition* from one to another. This transition is explained by the *development* of the productive forces. The development of the productive forces explains changes in the relations of production because it creates the conditions in which such changes occur. We can therefore identify the key role of the development of the productive forces in Marx's theory as the creation of conditions for revolutionary transformations of society. I will call this the *Revolutionary Effect* of productive development. The terms in which I have identified this role are, however, too vague to serve as anything more than a starting-point. In order to determine which forms of technological development are presupposed by this role, we must look more closely at *how* the development of the productive forces produces its Revolutionary Effect. Two elements of this effect can be discerned right away: the idea that the productive forces must reach a certain level of development to *make possible a new social form,* and the idea that at a certain level of development the productive forces *undermine the viability of the old form.* I call these the Enabling Effect and the Undermining Effect. Before examining these, however, it may be helpful to clarify some of the concepts that have just been introduced in relation to G. A. Cohen's functional interpretation of historical materialism, which will form the basis for the analysis that follows.

Marx, we have seen, thinks that the production relations existing in a society are explained by the level of development of the productive forces. Cohen interprets this claim as a species of functional explanation, which in

---

[32] 1859 Preface, p. 181; *The Poverty of Philosophy,* p. 92.

turn is a species of consequence explanation. In a consequence explanation a phenomenon (or property, process, etc.) is explained by its propensity to produce certain effects. Thus, according to Cohen's interpretation of Marx, 'production relations have the character they do because, in virtue of that character, they promote the development of the productive forces'.[33] In a functional explanation the effect that enables the phenomenon producing it to be explained in this way is a function of that phenomenon; thus, on Cohen's account, Marx's explanation is a functional one since *it is a function of the relations of production to promote the development of the productive forces*.[34]

This may appear, at first sight, to be a reversal of what I have called the Revolutionary Effect of productive force development, i.e. the fact that *the role of productive force development within Marx's theory is to bring about changes in the relations of production*. It may therefore appear that what we have is a symmetrical pattern of explanation in which the forces are explained by their effect on the relations and the relations are explained by their effect on the forces. It is questionable whether a functional explanation can have this form, but in any case a symmetrical explanation of the forces by their effect on the relations of production, and the relations by their effect on the forces of production, would be inconsistent with Marx's assertion of the explanatory primacy of the forces over the relations.[35]

In fact, the Revolutionary Effect that I ascribe to the development of the productive forces is consistent with Marx's account of the asymmetrical relation between forces and relations of production. This can be seen by looking again at the way functional explanation works. The fact that a phenomenon (or property, process, etc.) has a propensity to produce a certain type of effect can only explain its occurrence if there is a general tendency (or perhaps something stronger) for phenomena that produce effects of this type to occur. Cohen cites a standard example: 'Birds have hollow

---

[33] Cohen 1978, p. 249.

[34] The distinction between functional and consequence explanations is somewhat hazy, since there are competing accounts in the literature of what constitutes a function. Elster (1985, p. 27; Elster 1989, p. 50), for example, defines a function as a consequence that is beneficial to someone or something, or, within the social sciences, as a consequence that is beneficial to some dominant economic or political structure. For Larry Wright, on the other hand, for something to be a function of something else is for it to be a consequence of that thing which also explains that thing's existence (Wright 1973; see also Searle 1995, p, 16). This, it may be noted, collapses the distinction between consequence and functional explanations, since any consequence which features in a consequence explanation will, on this account, be a function. Cohen's characterisation of Marx's explanation as a functional one, however, holds on either of these accounts.

[35] The explanatory primacy of the forces of production in Marx's texts is argued for in Cohen 1978, ch. 6, and Cohen 1988, ch. 1.

bones because hollow bones facilitate flight.'[36] We find this acceptable as an explanation of the fact that birds have hollow bones, because we think that in general species tend to acquire characteristics that confer reproductive and survival advantages, and that the ability to fly is such a characteristic. Although we might assert this tendency on the grounds of observation alone,[37] we do in this case have a causal model – Darwin's theory of natural selection – which enables us to see why such a tendency exists.

Similarly, the statement that a particular set of relations of production facilitates development of the productive forces explains the existence of those relations only if it is the case that relations of production tend to emerge and persist which promote the development of the productive forces. Cohen thinks that we could have grounds for asserting a tendency of this kind without knowing the causal mechanism that gives rise to it,[38] but that, as in the biological case, an account of the underlying mechanism is in fact available. The tendency exists because (some) people benefit from the development of the productive forces and therefore tend to select relations which promote that development. The fact that development of the productive forces creates conditions in which it is rational to select new relations of production (i.e. its Revolutionary Effect) is thus a part of the causal mechanism underlying Marx's functional explanation of social change. In attributing the Revolutionary Effect to the development of the productive forces I am not giving a functional explanation of their development, since there is no suggestion that development of the productive forces occurs *because* it performs this function.[39] The Revolutionary Effect does, however, signal the importance of the development of the productive forces as an explanatory element within Marx's theory, and also raises that possibility that Marx – along with others who accept his explanation of social change and share his desire to be rid of capitalism – may be committed to the *desirability* of that development in some form and to some

---

[36] Cohen 1978, p. 249.

[37] Observation, that is, of living species and of the fossil record.

[38] As noted in chapter 3, Cohen's view here contrasts with that of Jon Elster, who argues that a functional explanation is only valid where the underlying causal mechanism is specified, and that where the underlying mechanism *is* specified the functional explanation is redundant. As argued in chapter 3, however, we need not accept this view: since (i) we can have empirical grounds for believing that a tendency exists without knowing why (i.e. without knowing the underlying causal mechanism), and (ii) there may be pragmatic reasons for explaining phenomena at the functional rather than the causal level, for example if the causal story is very complicated.

[39] Not usually, at least. A partial exception might be the rapid industrialisation of the USSR, promoted by its rulers with the expressed purpose of providing the proper conditions for socialism, and for the transition to the higher stage of communism.

extent. In what form, and to what extent, are questions that an analysis of the two components of the Revolutionary Effect should help us to answer.

5.4.1 The Undermining Effect

Let us return, then, to the two elements of the Revolutionary Effect, and their implications for the kinds of technological development that can constitute development of the productive forces within Marx's theory. The idea that productive forces can come into conflict with, and thus undermine, the existing relations of production is central to Marx's understanding of the historical process. This conflict arises, he argues, when the productive forces develop to a point where they are constrained or *fettered* by the relations of production. The concept of fettering is invoked in several of Marx and Engels's works, but it receives its classic exposition in a passage from the 1859 Preface:

> At a certain stage of their development, the material productive forces of society come in conflict with the existing relations of production, or – what is but a legal expression for the same thing – with the property relations within which they have been at work hitherto. From forms of development of the productive forces these relations turn into their fetters. Then begins an epoch of social revolution.[40]

Despite frequent use of the concept, however, Marx fails to provide a clear definition of what it is for the relations of production to fetter the productive forces, and divergent formulations in his writings have given rise to competing interpretations.

On one interpretation, fettering occurs when the relations of production prevent any further development of the productive forces. Following Miller and Cohen we may refer to this conception of fettering as 'Absolute Stagnation'.[41] This conception of fettering has been tacitly assumed by many commentators[42] and is implicit in some of Marx's formulations,[43] but is unsatisfactory as an explanation of social change. For one thing, the focus on absolute stagnation appears arbitrary: why should we assume

---

[40] 1859 Preface, pp. 181–2. Similar thoughts are expressed in *The German Ideology*, p. 87, *Manifesto of the Communist Party*, p. 40, and *Grundrisse*, p. 749.

[41] Miller 1981, pp. 96–7; Cohen 1988, p. 109. Elster (1985, p. 259) refers to this conception of fettering simply as 'Stagnation'. Cohen also uses the term 'Absolute Development Fettering' (p. 114); this, however, should be considered a broader term than 'Absolute Stagnation', since a conception of fettering as the slowing down of the development of the productive forces, considered in itself and not in comparison with any other actual or possible society (cf. Relative Development Fettering, discussed below), would also be a case of Absolute Development Fettering, as would McMurtry's conception of fettering in terms of 'absolute forfeiture' of the productive forces (discussed by Graham 1992, pp. 68–9).

[42] See Elster 1985, p. 260.

[43] For example, the passage from the 1859 Preface quoted below, text to note 60.

that it is precisely a *halting*, rather than a slowing down or regression in the development of the productive forces, that is needed in order to motivate the overthrow of existing relations of production? Moreover, the Absolute Stagnation conception of fettering fails what Cohen calls the 'predictability constraint'. This (along with the 'revolution constraint' discussed below) is one of two constraints on what can count as a satisfactory interpretation of fettering, derived by Cohen from Marx's account of fettering in the Preface.[44] According to the predictability constraint a satisfactory conception of fettering must be one whose occurrence in the future can plausibly be anticipated. Absolute Stagnation fails to satisfy this condition since, as Cohen argues, we have no reason to think 'that, were capitalism, for example, to last forever, then the development of the productive forces would at some point entirely cease'.[45]

An alternative interpretation is 'Relative Development Fettering', also referred to as 'Relative Inferiority' and 'Suboptimality'.[46] On this account the productive forces are fettered by the relations of production when the productive forces develop more slowly than they would under an alternative set of relations. This conception is weaker than Absolute Stagnation and may therefore pass the predictability constraint; however, it fails what Cohen calls the 'revolution constraint'. The revolution constraint states that, in order to fulfil the explanatory role ascribed to fettering by Marx, 'it must be plausible to suppose that when relations become fetters they are revolutionized'.[47] Relative Development Fettering fails this test, according to Cohen, since 'the costs and dangers of revolution . . . make it unreasonable to expect a society to undergo revolution just because relations which are better at developing the productive forces are possible, especially when those relations have not already been formed elsewhere and been seen to be better'.[48] The point is not just that the benefits of revolution are uncertain in the absence of prior historical experience, but also that the benefits, even if real, may be insufficient to motivate revolutionary action, since Relative Development Fettering does not imply any impoverishment of the population and indeed is consistent with accelerating development of the productive forces. 'Is it plausible', Cohen asks, 'to suppose that revolution would be risked at a time of *accelerated* development of the productive forces, just because there would be still faster development under differ-

---

[44] Cohen 1988, pp. 109–10.   [45] *Ibid.*, p. 110.

[46] Cohen introduces the term 'Relative Development Fettering' at p. 114, though he mostly follows Miller's (1981, pp. 96–7) term, 'Relative Inferiority'. The term 'Suboptimality' is from Elster 1985, p. 259. I have adopted the first term since the others are ambiguous between Development and Use Fettering.   [47] Cohen 1988, p. 110.

[48] *Ibid.*, p. 111; see also Elster 1985, p. 293; Graham 1992, p. 70.

ent relations? Would workers overthrow a capitalism which has reduced the length of each computer generation to one year because socialism promises to make it nine months?'[49]

These problems may to an extent be addressed by interpreting 'fettering' as applying not to the *development* of the productive forces but to their *use*. The suggestion that fettering involves a limit not on a society's development of new productive forces but on its ability to use highly developed forces finds support in a number of passages from Marx. For example, in the *Communist Manifesto* Marx writes of feudal relations of production having become fetters on 'the *already developed* productive forces', and of capitalist fetters having been *overcome* by the development of the productive forces. The latter is evidenced for Marx not by a slow-down in the development of the productive forces but by the periodic crises of overproduction in which overdevelopment of the productive forces relative to capitalist relations leads to 'a state of momentary barbarism' in which 'it appears as if a famine, a universal war of devastation had cut off the supply of every means of subsistence'.[50]

Use Fettering, like Development Fettering, can be defined in absolute or relative terms. However, since no version of Absolute Use Fettering is remotely plausible[51] we can concentrate on Relative Use Fettering – the circumstance that a particular set of production relations uses its productive forces less effectively than some alternative set of relations. This conception of fettering plausibly satisfies Cohen's predictability constraint, given the reasonable assumption that relations of production are only finitely flexible in the range of productive forces that they can efficiently put to use.[52] It also has the advantage of being more easily and concretely perceptible than Relative Development Fettering, and therefore more plausible as a motivation for revolutionary action. As Graham argues:

If productive forces are already visible whose use would benefit subordinate groups, there is a plausibility in the hypothesis that they will experience frustration and attempt to create circumstances where the forces can be used and the

---

[49] Cohen 1988, p. 111.

[50] *Manifesto of the Communist Party*, pp. 40–1. See also the passages cited in Cohen 1988, p. 116, Elster 1985, pp. 264–5 and Graham 1992, p. 69, especially the claim, from *The German Ideology*, p. 78, that 'a great multitude' of the forces developed under capitalism 'could find no application at all within this system'.

[51] Cohen, for example, defines (but rejects as even less plausible than Absolute Development Fettering) a version of Absolute Use Fettering according to which none of the available productive capacity is used. An alternative conception of Absolute Use Fettering would characterise it as a failure to utilise any *new* productive forces. But while such a conception might find support in the *Communist Manifesto* account, discussed above, it too would fail Cohen's predictability constraint.   [52] Cf. Graham 1992, p. 76; Cohen 1988, p. 114.

benefit received. There is less plausibility in hypothesizing that this will happen where such productive forces have not even been developed.[53]

Despite these advantages, however, there is a difficulty with Relative Use Fettering. A society which uses its productive forces inefficiently may nevertheless develop those forces faster than another society which uses its forces more efficiently. (Efficiency here refers to the proportion of available productive forces that are actually used.) If such circumstances persist then the society in which productive forces develop quickly but are used inefficiently will quickly overtake the efficient but slow-developing society, in terms of its net *used productive power* (the result of productive capacity and proportion of productive capacity used). It has been suggested that a circumstance of this kind may exist in relation to capitalism and communism, the latter being less wasteful in its use of existing productive capacities but less good at developing new ones.[54] If what matters to the inhabitants of a society is its used productive power, then the fact that capitalism fettered the use of productive forces would not, in such circumstances, be a reason for overthrowing it. In response to considerations of this kind Cohen proposes another conception of fettering, 'Net Fettering', according to which relations of production fetter if the used productive power resulting from the level of development and rate of use of the productive forces is less than under some alternative relations.[55] This conception of fettering deals with the fact that a high rate of development of the productive forces could compensate for (or outweigh) a low rate of use; it does not, however, take account of Graham's point, that the non-utilisation of existing productive forces is more visible and a more plausible source of revolutionary motivation than the potential for developing productive forces that do not yet exist. What this might suggest is that people have reason to overthrow existing relations of production if the used productive capacity under those relations follows a lower trajectory than under some alternative relations, but that this is only likely to become an actual source of motivation if the lower trajectory is to a significant extent a result of existing but unused potential.[56]

---

[53] Graham 1992, p. 150.   [54] Elster 1985, p. 266, citing Schumpeter.

[55] Cohen 1988, p. 117. Strictly speaking, of course, the definition I have given is of *Relative* Net Fettering. See also Elster 1985, p. 267, where it is asserted that the growth of net social product is the relevant criterion for comparing social systems.

[56] In practice this would mean that existing relations would have to be inferior to the alternative in terms of both development and use of the productive forces to provide rational revolutionary motivation. A related suggestion, considered and rejected by Elster (1985, p. 294), is that Use Fettering might provide the motivation in the first country to undergo a revolutionary transition, but that revolutions in subsequent countries might be motivated also by Development Fettering, once the first country has demonstrated its developmental superiority.

It is arguable, however, that the attempt to combine Development Fettering and Use Fettering within the single metric of Net Fettering is mistaken, since Use Fettering (at least as it is manifested by capitalism) is best treated as a qualitative notion. The problem with capitalism's use of productive forces, on this account, is not that too small a proportion is used, but rather that they are used in the wrong ways, or for the wrong purposes. As Graham puts it, they are used to increase surplus value, with only incidental benefit to human beings; '[t]he claim must be that capitalism dictates a less rational use than would be possible in some alternative system'.[57] Once we interpret Use Fettering in this qualitative way it is no longer true that its effect can be offset by faster development of the productive forces. (If you are driving in the wrong direction, doing so in a faster car is likely to make matters worse, not better.) Development Fettering may also be understood qualitatively, as Marx himself does when he writes, in the *German Ideology*, that the productive forces associated with large-scale industry 'received under the system of private property a one-sided development only, and became for the majority destructive forces'.[58] Understanding fettering in these qualitative terms also increases the likelihood of Use Fettering and Development Fettering coinciding, since a society which fails to use its existing productive forces for certain purposes is unlikely to develop them very extensively in the ways appropriate to those objectives.

The question to be addressed here is whether the Undermining Effect of productive force development, elaborated in terms of fettering, commits Marx to the desirability of ecologically damaging forms of that development. There are two reasons why this might appear to be the case.

First, in claiming that capitalist relations of production are undermined by their fettering of the development or use of productive forces, Marx is assuming that people (at least those who do not have an overriding interest in maintaining the status quo) have an interest in the unfettered development and use of the productive forces and therefore a reason to overthrow the fettering relations. Marx clearly identifies himself with this perspective, regarding the fettering of productive forces by capitalism and their unfettered development under communism as a reason for preferring the latter. However, since the motivating interest that Marx ascribes to people is an interest in their own welfare, and since his own normative stance is informed by a concern for human welfare more generally, there is no reason to ascribe to him the view that communism will or should engage in indiscriminate or ecologically damaging developments or uses

---

[57] Graham 1992, p. 152.   [58] *The German Ideology*, p. 78.

of productive technology. Indeed the need to discriminate between technologies that enhance welfare and those that diminish it is implied by Marx's own criticism of the 'one-sided' and 'destructive' ways in which productive forces are developed and used under capitalism.

Second, development of the productive forces is necessary, for Marx, in order for them to reach the level at which they become fettered by capitalism, creating the conditions for its overthrow. This appears ecologically more problematical, since the development in question must take place under capitalism and will therefore, according to Marx, be shaped by considerations of profit rather than human welfare. The question, therefore, is whether Marxists are committed to supporting such development, irrespective of its ecological and social consequences, and despite green opposition, as a necessary prerequisite for the overthrow of capitalism.

In the *Communist Manifesto*, Marx and Engels avoid any such commitment by insisting that the productive forces have already reached the stage at which they are fettered by capitalist relations of production. After describing how feudal relations became fetters upon the development of the productive forces and were 'burst asunder', they assert that '[a] similar movement is going on before our own eyes', and that '[f]or many a decade past the history of industry and commerce is but the history of the revolt of modern productive forces against modern conditions of production, against the property relations that are the conditions for the existence of the bourgeoisie and of its rule'.[59] If, as this implies, capitalism had already become a fetter by 1848, then Marx has no reason (at least as far as fettering is concerned) to regard continued capitalist development of the productive forces as desirable. However, given the worldwide dominance and continued productive growth of capitalism some century and a half later, could contemporary Marxists be committed to the view that fettering has yet to take place and that any productive developments must be supported, whatever their consequences, as necessary steps towards the time when capitalism will become a fetter and be 'burst asunder'? Marx himself appears to move towards such a view in the 1859 Preface, where, in a comment perhaps intended to account for the failure of the revolutions of 1848 to initiate the socialist transformation predicted in the *Manifesto*, he writes that '[n]o social order ever perishes before all the productive forces for which there is room in it have developed'.[60]

However, neither the persistence of capitalism nor its continued technological innovation requires the conclusion that further development of the

[59] *Manifesto of the Communist Party*, p. 40.   [60] 1859 Preface, p. 182.

productive forces is necessary before fettering takes place. The continued development of the productive forces under capitalism is consistent with it being a fetter on any interpretation of fettering except Absolute Stagnation, and while Marx appears to endorse this interpretation in the passage just quoted, we have already seen that it is an implausible interpretation on theoretical grounds and that other passages in Marx support different interpretations. Moreover, there is no need to suppose that fettering has yet to take place in order to account for the persistence of capitalism, since its persistence *post*-fettering may be accommodated within Marx's theory if the fettering of the productive forces is considered to be a necessary but not a sufficient condition for a change in the relations of production. It might for example be that the level of productive development is sufficient to perform the Undermining Effect but not yet sufficient to perform the Enabling Effect (to be discussed in the next subsection), resulting in crises and instability but no viable alternative. This would be consistent with Marx's claim that the fettering of the productive forces ushers in an *epoch* of social revolution.

There are, in addition, reasons for thinking that an adequate account of fettering cannot require Marxists to support ecologically damaging productive developments as steps towards its realisation. The important point about fettering, illustrated by Cohen's revolution constraint, is that fettering occurs when productive development or use is constrained in such a way as to give people (those who are members of the revolutionary class, at least) reason to replace the fettering relations of production with others which remove that constraint. The question then is: under what sorts of circumstances do people have such a reason?

Our earlier discussion indicated that technological innovation in the sphere of production may be pursued for a variety of reasons, which may from different perspectives be regarded as criteria of technological development. For example, technological innovation may be used to increase the quantity or improve the quality of a product, to reduce the labour input or monetary costs of producing it, or to reduce the ecological impact of its production in terms of resource use and pollution levels. The next step is to recognise that any reason people have for pursuing technological development may also be a reason for removing constraints upon that development. It follows that in principle fettering may occur when the development or use of productive forces in accordance with any one of these objectives is hindered by the prevailing relations of production, although in practice each agent's interest in technological development is likely to involve several of these objectives, and constraints upon one form of development may be offset by development of another element of the bundle. There will, however, be limits to such trade-offs, particularly where people's most basic needs are concerned. Relations of production which allow productive technology to develop, but not in the ways that are required in order to mitigate its ecological impact, should therefore be counted as fetters as soon as the detrimental impact of this constricted form of development gives sufficient reason to abolish those relations. The Undermining Effect therefore cannot require that such developments be tolerated as means to bringing about fettering, since a society that has room only for developments of this kind already acts as a fetter.

5.4.2 The Enabling Effect

Fettering, whether of the development or the use of the productive forces, cannot be sufficient reason for overthrowing existing relations of production unless there is some alternative that is both better and feasible. In other words, the Undermining Effect of the development of the productive forces is only revolutionary if some such alternative exists. Which alternative relations are viable at any time depends, for Marx, upon the level of development of the productive forces. The effect that the development of the productive forces has in making new relations of production viable is what I call its Enabling Effect. It may appear that if we interpret the Undermining Effect in terms of Relative Fettering, then the Undermining Effect and the Enabling Effect will amount to the same thing, namely the superiority of the new relations of production over the old. However, although there is some overlap, Marx's account of the Enabling Effect is richer than such an account would imply, and the conditions that make possible a new set of relations are not exhausted by such a comparison.

Although Marx does not provide any general summary of the Enabling Effect of productive development comparable with his account of its Undermining Effect in the Preface, it too is integral to the theory of historical materialism. The idea that each set of relations of production becomes viable only when a certain minimum level of productive development is reached helps to explain Marx's conviction that society must pass through a succession of different relations of production before socialism can emerge.[61] Each new set of relations is in its turn made possible by a development of the productive forces which progressively increases the quan-

---

[61] This is not the only means Marx has at his disposal to explain the necessity of stage-by-stage social development, since he could also appeal to the idea that each set of relations can only emerge from a particular predecessor, in which some class has the motivation and

tity of goods produced beyond what is required to sustain the lives of the producers. As Cohen summarises it:

At the first stage, productive power is too meagre to enable a class of non-producers to live off the labour of producers. The *material* position is one of absence of surplus, and the corresponding *social* (or *economic*) form is a primitive classless society.

In the second stage of material development, a surplus appears, of a size sufficient to support an exploiting class, but not large enough to sustain a capitalist accumulation process. The corresponding social form is, accordingly, a pre-capitalist class society . . .

At stage 3 the surplus has become generous enough to make capitalism possible.[62]

Similar conditions apply for the disappearance of class society. Marx is clearly committed to the proposition that in a socialist society the increased productivity of labour made possible by the advance of technology should be used to reduce the burden of labour,[63] yet he is equally clear that an increased level of output must be achieved. In the *German Ideology* Marx and Engels write that for communism to emerge successfully, 'a great increase in productive power, a high degree of its development . . . is an absolutely necessary practical premise because without it *want* is merely made general, and with *destitution* the struggle for necessities and all the old filthy business would necessarily be reproduced'.[64]

And similarly, in the *Critique of the Gotha Programme*, Marx writes that the 'higher phase of communist society' – characterised by its distribution according to needs – can be entered only 'after the productive forces have . . . increased with the all-round development of the individual, and all the springs of co-operative wealth flow more abundantly'.[65] Since technological development that is oriented towards the production of more goods has the potential to be ecologically damaging, it will be necessary to look more closely at these preconditions. However, there is nothing in the passages quoted above to imply that Marx is committed to an *unceasing* rise in output, and his commitment to reducing the burden of labour provides a

---

ability to bring it about. However, it should be apparent from what follows that the successive 'enabling' of relations of production by the development of the productive forces forms at least part of Marx's explanation of their sequential emergence.

[62] Cohen 1988, pp. 155–6. The dependence of the emergence of exploitative societies on the growth of production is noted by Engels in *The Origin of the Family, Private Property and the State*, pp. 568–9. The accumulative character of capitalism is discussed in *Capital*, vol. I, part 7.

[63] See, for example, *Capital*, vol. I, ch. 10, especially section 1 on 'The Limits of the Working Day', and ch. 15, especially section 3(b) on 'The Prolongation of the Working Day'; also *Grundrisse*, pp. 701, 708; and *Capital*, vol. III, p. 959, on the reduction of the working day as the basic prerequisite for creating a true realm of freedom.

[64] *The German Ideology*, p. 56. See also p. 107.   [65] *Critique of the Gotha Programme*, p. 320.

reason for stabilising output once it is sufficient for the needs of a communist society. This would limit the ecological consequences of the development and allow the possibility of these consequences being offset by improvements in the ecological efficiency of productive technology. Indeed, if the reason communism requires productive development is to permit the meeting of human needs, then improvements in the ecological efficiency of productive technology may be deemed an essential part of that development, since the ecological consequences that would otherwise ensue would pose a threat to those needs.

This last point indicates that the Enabling Effect of productive development may have a qualitative as well as a quantitative dimension. The suggestion here is that what is required in order to make communism possible is not just a general increase in the power of technology, but rather the development of particular kinds of technology. This may appear to be a departure from the way in which Marx writes of the development of the productive forces as a precondition for communism, but it has parallels in his account of the emergence of capitalism. Marx and Engels may be interpreted as attributing a qualitative role of this kind to transport technology, when, in the *Communist Manifesto* and elsewhere, they describe how improvements in navigation and shipping facilitated the growth of merchant capital and the plunder of natural and human resources in the colonies, which in turn allowed industrial capital to become the dominant form of production:

The discovery of America, the rounding of the Cape, opened up fresh grounds for the rising bourgeoisie. The East-Indian and Chinese markets, the colonisation of America, trade with the colonies, the increase in the means of exchange and in commodities generally, gave to commerce, to navigation, to industry, an impulse never before known, and thereby, to the revolutionary element in the tottering feudal society, a rapid development.[66]

My purpose in quoting this passage is simply to show that the idea of a qualitative dimension to the Enabling Effect of productive development is not alien to Marx's own account. The suggestion above was that another kind of qualitative development – an increase in the 'ecological efficiency' of productive technology – may be presupposed as a condition for communism. This may appear to be in tension with the earlier suggestion that the *fettering* of such beneficial developments, and the misdirection of pro-

---

[66] *Manifesto of the Communist Party*, p. 36. See also 'Results of the Immediate Process of Production', p. 1023, on the role of merchants' capital as 'the soil from which modern capitalism has grown', and *Capital*, vol. I, p. 915, on the discovery and exploitation of America and Africa as 'the chief moments of primitive accumulation'.

ductive force development and use towards more damaging ends, may contribute to the Undermining Effect of productive force development. The point, however, is that the viability of communism may depend upon technology offering the *potential* to achieve a reasonable satisfaction of human needs whilst averting ecological problems that would undermine this objective. That potential may, however, be unrealised under capitalism, because of both inappropriate use and constraints on the ways in which technologies are developed.

In itself the inclusion of an enhanced potential for ecological efficiency within the technological prerequisites for communism suggests that these prerequisites are likely to be ecologically benign. A problem arises, however, when we consider this alongside other qualitative developments of productive technology that might be presupposed by Marx's communist project. For example, it has been suggested that one such precondition is the development of technologies that will make productive activity a less alienating experience than it is with current productive technology.[67] The problem, then, is whether a development is possible which satisfies *all* of these preconditions; we may wonder, for example, whether the technological conditions for eliminating alienation are compatible with those for reducing ecological impact. For now I will just note this as a problem to be returned to later.

5.5 Explaining productive development: an autonomous tendency?

I have argued (with the proviso just noted) that the Revolutionary Effect attributed by Marx to the development of the productive forces may in principle be produced by ecologically benign forms of that development, and that Marxists are not, therefore, committed to viewing ecologically damaging forms of technological development as a price that must be paid to bring about circumstances in which a revolutionary transformation of society can occur. In one respect, however, this argument is incomplete. Marxism claims to provide an account of how a revolutionary transformation of existing society can take place. It is therefore not enough to show

---

[67] This has been suggested by Grundmann 1991b. See also Naletov's (1984, pp. 370–3) suggestion that the development of ergonomics, as the science of 'the optimal interaction between man and technical means', is necessary to overcome the alienation and rigid division of labour imposed by contemporary productive technology. Another kind of technological development that might be thought necessary for the successful implementation of Marx's communist project is the recent and ongoing revolution in information technology, which might be thought important as a means of overcoming the unwieldiness and co-ordination problems that have characterised planned economies in the past.

that a hypothetically postulated development of ecologically benign technologies *could* undermine the viability of a capitalism which failed effectively to implement and further develop this potential, and *could* provide the technological prerequisites for a viable communist society. In order to justify Marx's belief in the possibility of a communist future it is also necessary that the proposed conception of productive force development is one which we have reason to think *will* occur and will continue (provided capitalist fetters are removed at the appropriate stage) as far into the future as is necessary to create the conditions in which that society can develop and mature.[68] The question then arises of whether the mechanism postulated to explain such future development is consistent with that development taking the ecologically benign forms hypothesised above.

One account, which initially appears unpromising in this respect, and therefore raises in stark form the problem to be addressed in this section, is Cohen's influential account of an 'autonomous tendency for the productive forces to develop'. The problem for Cohen's account is that in characterising the tendency of the productive forces to develop as 'autonomous' it may appear to render that development insensitive to changes in circumstances such as the emergence of ecological problems, and to dictate a more historically uniform, or unilinear, interpretation of it than the preceding argument suggests.[69]

Cohen defines the autonomy of this tendency as 'its independence of social structure, its rootedness in fundamental material facts of human nature and the human situation'.[70] Marx, Cohen believes, is committed to this autonomy by his central claim that '[t]he nature of the production relations of a society is explained by the level of development of its productive forces'.[71] According to Cohen, Marx's claim that the development of the forces explains the development of the relations is an assertion of the *explanatory primacy* of the former. The claim is that 'the nature of a set of production relations is explained by the level of development of the productive forces embraced by it (*to a far greater extent than vice versa*)'.[72] This asymmetry demands that the development of the productive forces be

---

[68] This requirement parallels the 'predictability constraint' that Cohen places upon the interpretation of fettering.

[69] What my argument below indicates, however, is: (i) that Cohen's account gives a more significant role to production relations than his critics sometimes realise (as shown by Cohen's reply to Levine and Wright); and (ii) that when the suggested amendments, motivated by Cohen's own argument, are incorporated, this yields an account of productive development which has the potential to be sensitive to a wider range of material circumstances than Cohen realises, including the emergence and development of ecological problems.   [70] Cohen 1988, p. 84.   [71] Cohen 1988, p. 84; Cohen 1978, p. 134.

[72] Cohen 1978, p. 134; my emphasis.

explained, at least in part, by something other than the relations of production. Additionally, if the applicability of narrow historical materialism across different epochs is not to be regarded as some sort of historical accident, an explanation must be given of why the productive forces tend to develop within all the various social structures to which that development gives rise.

For Cohen, the tendency of the productive forces to develop arises from the conjunction of three (asocial) facts, one about the human situation and two about human nature. Firstly, humans live in a situation of material scarcity; secondly, they have the capacity to devise more powerful productive forces; and thirdly, they are rational enough to grasp the opportunities provided by this capacity to ameliorate the scarcity under which they labour. Given these facts, Cohen argues, 'productive power will . . . tend, if not always continuously, then at least sporadically, to expand'.[73]

The trouble with this is that it seems unlikely that a developmental tendency based on unchanging facts about human nature and the human situation could undergo the required shift from the forms of technological development that have increased productive output to its present levels but at great ecological cost, to forms which could play a part in reducing those costs. However, what Cohen does not say is that, important though these unchanging facts are, they only partially describe the material conditions under which humans exercise their inventive capacities and make decisions about their productive activities. For Cohen the chief problem faced by humans is having to spend a large proportion of their time producing to meet their needs. Their ingenuity, according to Cohen, enables them to come up with solutions to this problem in the form of technological innovations to increase labour productivity, and their rationality ensures that the best solutions are adopted and retained. However, as was indicated in the discussion of Jarvie, humans also face a range of problems arising out of the concrete social and material circumstances in which they find themselves. Acknowledging that these problems too may motivate the use of our innovative and rational capacities allows us to explain and predict the occurrence of new and varied forms of technological development including, for example, development aimed at reducing the quantity of scarce raw materials that must be consumed in order to meet our needs, or at reducing the deleterious unintended consequences of our need-meeting activity. And, given that the ways in which ecological problems affect people depend on their position within the social structure, this

[73] Cohen 1988, p. 86.

account also makes it plausible to suppose that the extent to which these ecologically oriented forms of technological development feature in the development of the productive forces will depend upon the structure of interests within the prevailing relations of production.

This last point echoes the emphasis of much of Marx's writing on the historical specificity of different social structures, and offers the prospect that a change in the relations of production may facilitate a more benign development of productive forces than under capitalism. It suggests, however, that, since we have now included among the motivations for productive development not only the unchanging facts of human existence but also the more specific forms in which humans may experience the effects of material scarcity, the tendency of the productive forces to develop is no longer, as Cohen claims, autonomous of social structure. To investigate this, let us consider Cohen's response to another argument against the autonomy of productive force development, put forward by Andrew Levine and Erik Olin Wright.[74]

Levine and Wright argue that since people's relation to the productive forces, and therefore their interest in productive development, depend upon the social structure within which they live and the position they occupy within it, the tendency for the productive forces to develop cannot be autonomous of social structure but must be explained by a succession of 'class-specific rationalities'[75] corresponding to successive social structures. For example, they argue that the development of the productive forces under feudalism cannot be explained as the rational action of the direct producers seeking to reduce their workload, since the economic structure of feudalism is such that the direct producers – peasants – do not benefit from the development of the productive forces; rather, Levine and Wright contend, the development of the productive forces under feudalism must be explained by the interests of the feudal lords, determined by *their* position in the economic structure. Similarly, we may observe that development of the productive forces under capitalism cannot be explained by the interest of workers in reducing their labour time, since (i) it is not generally in the workers' power to alter the productive forces that they use, and (ii) they rarely benefit from developments of the productive forces; often their working hours rise even as labour productivity

---

[74] Levine and Wright 1980.

[75] The term 'class-specific rationalities' is misleading because it is not actually the rationality which is class-specific, but the conditions under which it is exercised. Levine and Wright's model in fact assumes that the reasoning of all classes conforms to a standard conception of instrumental rationality as valid means–end reasoning. What may differ between different classes is the set of ends pursued, and what must be done in order to achieve any particular end.

increases, or else they find themselves unemployed. Rather, as Marx shows in *Capital*, development of the productive forces under capitalism is to be explained by the interest of capitalists not in reducing workloads but in increasing per capita production in order to produce more surplus value. Thus, according to Levine and Wright, it is not universal human reason and ingenuity in the face of scarcity which brings about development of the productive forces, but sets of class interests such as those of the feudal lord or the capitalist, interests which are specific to particular social forms.

Cohen responds that Levine and Wright have mislocated the point of action of rationality within his model. According to Cohen's model it is indeed the class-specific rationality in operation at any given time which is the *immediate* cause of the development of the productive forces; but the autonomy of this process, its independence of social structure, is ensured by the operation of universal human rationality at a higher level. Because of their rationality and because they live in circumstances of scarcity people select just those structures which promote and do not fetter the development of the productive forces, and it follows from this that the existence of a developmental tendency is independent of *which* structure and *which* class-specific rationality is in operation.[76]

So, on Cohen's account, the selection of social structures underlies the tendency of the productive forces to develop and gives that tendency its autonomy. Cohen's assumption is that the motivation for the selection of structures is always the pursuit of improved labour productivity. I have argued, however, in the discussion of the Undermining Effect above, that the selection of structures may be motivated by various criteria of technological development, including ecological considerations, and that some elements of this motivation, concern about the ecological effects of production being one of them, are not universal but arise from particular circumstances.[77] Since, for Cohen, it is the selection of social structures that gives

---

[76] Cohen (1988, pp. 89–90) expresses this by saying that although there is an *autonomous tendency* for the productive forces to develop (since the tendency exists independently of which social structure obtains), there is not a tendency for the forces to *develop autonomously* (since productive forces only develop in the context of suitable relations of production). An alternative way of expressing this would be to say that there is a *transcendental* tendency for the productive forces to develop. This terminology expresses the idea that although social structure is involved causally in the development of the productive forces, the tendency is universal in virtue of the fact that it is a condition for the existence of any structure that it produces such a tendency.

[77] It could be argued that ecological problems have the same source in the universal condition of material scarcity as the need to spend large amounts of time working, but so far as the selection of social structures is concerned the point is that whereas (if Cohen is to be believed) this scarcity always manifests itself in the latter way, it is only manifest in ecological problems when the productive forces become powerful enough and the population large enough to exhaust resources, cause serious pollution, etc.

the productive forces their tendency to develop, it follows that this tendency may include whatever forms of development feature in the selection of structures. Contrary to initial appearances, then, the explanation of productive development offered by Cohen, when broadened in the way I have suggested, does not commit Marx to an ecologically insensitive conception of productive development. I will leave aside the terminological question of whether the tendency of the productive forces to develop, as we have now described it, is properly called an *autonomous* one. Suffice it to say that it satisfies the asymmetry requirement set out above, and that if it is an autonomous tendency, then this is an autonomy which allows for the development of the productive forces to take different forms under different relations of production.

There is, however, a further objection to Cohen's autonomy thesis that is not answered by his reply to Levine and Wright. The idea that social structures are *chosen* by individuals on the basis of their tendency to promote the development of the productive forces is rejected as implausible by Robert Brenner. Writing of the transition from feudalism to capitalism, Brenner says:

> One could not, I think, conceivably argue that individual pre-capitalist economic actors would seek to separate the pre-capitalist producers from their means of subsistence and break up the lord's institutionalized relationships with the producers which allowed them to extract a surplus by extra-economic compulsion in order to install a system where the individual actors had, as their rule for reproduction, the maximisation of profits . . . especially when such a system had never previously existed.[78]

In rejecting the view that social structures are selected for their technological dynamism Brenner is challenging Cohen's claim to have established an autonomous, transhistorical tendency for the productive forces to develop; for, he argues, 'only if we could conceive of the economic actors as making such an unlikely move, could we accept the theory that the growth of the productive forces was primary in the march of history'.[79]

As a historian, Brenner's focus is on the transition from feudalism to capitalism, and accordingly we can set aside some elements of his critique as secondary to our main concerns. It may be, for example, that feudal economic actors were prevented from making the sort of choice ascribed to them by Cohen by factors specific to the period in which they lived, such as their lack of education, the poor means of communication available to them, and the absence of modern social sciences capable of predicting the

---

[78] Brenner 1986, pp. 46–8, n. 13.   [79] *Ibid.*

effects of different social structures. Such factors would, however, have no bearing upon the ability of workers or others in contemporary capitalist societies to make such choices, and to do so *inter alia* on the basis of ecological considerations. Since it is this future transition and its effects upon the development of the productive forces which are of chief interest from an ecological perspective, we could also set aside Brenner's claim that feudalism is technologically stagnant[80] and concentrate instead on the effects of technological development under capitalism, the existence of which is not in dispute. It is, however, necessary to address the question of choice in the selection of relations of production, since, even were we to abandon claims to the transhistorical applicability of historical materialism, and Cohen's claim that the theory is based upon the autonomous tendency of the productive forces to develop, we would still need to account for the prospective transition from capitalism to socialism, a transition which has been characterised in terms of a choice of non-fettering relations of production over fettering ones.

Leaving aside considerations specific to feudalism, the main reasons for Brenner's rejection of the view that new relations of production are intentionally chosen by individuals on the basis of their tendency to promote development of the productive forces would appear to be: (i) that this implies an implausibly sophisticated understanding of social structures and their effects upon productive development on the part of members of the revolutionary classes, and (ii) that even if such an understanding were present there may be a gap between what is rational in the light of the transhistorical considerations given by Cohen and what is rational given the actual locations and socially conditioned interests of the agents to whom the choice is attributed.

One way of responding to this critique is to deny that the theory requires new relations of production to be intentionally selected. Functional explanation requires a selection mechanism favouring structures which produce certain effects, but does not require that those effects be intended by anyone. (Consider the explanation of birds' hollow bones referred to above.) Alan Carling and Chris Bertram both offer accounts of historical materialism in which international competition serves as the selection

---

[80] Brenner holds that only under capitalism do economic actors have a motive (market competition) that will lead to a systematic development of the productive forces (*ibid.*, p. 34). However, he acknowledges that technical progress may be expected in varying degrees even in pre-capitalist societies (*ibid.*, p. 41; cf. Carling 1991, p. 67 n. 4). This concession may be enough for Cohen, given that his autonomy thesis insists only on a sporadic growth of the productive forces and a preponderance of growth over regression (Cohen 1988, pp. 86, 99).

mechanism for new, more productive relations of production.[81] New relations of production may initially emerge for a variety of reasons, but those that lead to accelerated development of the productive forces will tend to persist owing to the economic and military advantage that they confer. Societies with less productive relations are likely either to have the new relations imposed upon them or to suffer internal crises leading to a change of relations. Whatever the general merits of such an account, however, it faces two objections as an account of the transition from capitalism to socialism. First, the idea that new relations may be imposed by economic or military domination, and even that they may be adopted internally in order better to resist such domination, conflicts with Marx's own insistence on *self-emancipation* as a necessary condition for the achievement of communism. One reason why Marx thinks self-emancipation necessary is that it is through collective activity aimed at overthrowing contemporary society that people can acquire the qualities necessary for the future society and form a collective vision of how they want that society to be.[82] This may be thought particularly pertinent if the new society is to be one in which ecological considerations play a heightened role not only institutionally but also internalised in people's conceptions of their own interests. Second, and relatedly, selection of social structures by means of economic and military competition may well favour rapid development of the productive forces but is unlikely to produce a shift towards more ecologically sensitive forms of that development. Military power is not known to depend upon ecological sensitivity, and economic competition, as presently constituted, tends to undermine the efforts of individual countries to maintain ecological standards.[83] New structures may to a degree be able to insulate themselves from market pressures, but are likely to face internal pressures if, for example, they provide fewer consumer goods for their population than other societies.[84] What this suggests is, first, that Marx may be right in suggesting that such fundamental change can only happen in several major countries at once (in order to avoid the expansion of consumer appetites that may result from cross-border comparisons),[85] and second, that if inter-societal competition

---

[81] Carling 1991, pp. 54–5; Bertram 1990. A similar argument is tentatively outlined in Cohen 1988, pp. 27–9. For a critique of Cohen's and Bertram's arguments, see Casal 1994.

[82] Graham 1992, pp. 135–8.   [83] Bertram himself (1990, p. 124) makes a similar point.

[84] There is an obvious parallel here with the dissatisfaction caused by a shortage of consumer goods in the former Eastern Bloc states.

[85] Marx asserts the necessity for revolution simultaneously in the major industrial countries in *Manifesto of the Communist Party*, p. 51. He notes the inflationary effect that comparison with others may have upon people's appetites in *Wage Labour and Capital*, p. 83. See also the discussion of false needs in section 6.2 below.

is to play a part in instigating new and more ecologically sensitive social structures then the very terms of that competition must be contested. The new structures must, in other words, arise from a conscious *choice* in favour of ecological protection over unlimited consumption, and so Brenner's scepticism about the possibility of such choices cannot be evaded.

A partial reply to Brenner's scepticism is already implicit in the discussion above of fettering. Brenner doubts whether people could have sufficient reason for believing in advance both that a new structure would produce enhanced development of the productive forces and that they would benefit from that development. It was argued above, however, that a motivation for structural change may more plausibly be constructed on the model of Qualitative Use Fettering. This eases the epistemic difficulties in the way of such motivation, since the potential to benefit from productive forces that already exist but are not appropriately used may be more readily perceptible than the potential to benefit from as yet unrealised future developments. Selection on this basis would produce a tendency for the productive forces to develop in 'appropriate' ways given the assumption stated above, that a society which uses productive forces for a certain purpose is more likely to develop them in ways appropriate to that purpose.[86] Brenner's second objection – concerning the divergence between what it is rational to do in the light of transhistorical human interests and what it is rational for real, socially situated actors to do – applies primarily to earlier transitions (including that from feudalism to capitalism which forms Brenner's focus), where the new relations to be instituted are exploitative ones in which benefits will accrue to a minority class at the expense of the majority. By contrast, the post-capitalist society envisaged by Marx is one in which exploitation will be eliminated and the development of the productive forces will be pursued for the benefit of humanity generally. It is therefore, in intention, a society the existence of which is in the interest of a large portion of the population.

None of this implies that the consciousness necessary to motivate the choice of new relations of production will arise automatically or inevitably. Marx himself was well aware of the ideological obstacles to such consciousness, seeing his own work as a practical intervention intended to overcome

---

[86] If this explanation of change is confined to the transition from capitalism to socialism then some other explanation would have to be given to justify Cohen's claim that there is a general tendency for the productive forces to develop in pre-socialist relations of production. For example, Graham (1992, p. 76) suggests that the development of the productive forces within successive structures up to and including capitalism is explained not by transhistorical principles of rationality, as Cohen asserts, but by the general interest of members of exploiting classes in compelling 'hard work with efficient technology from the subordinate class'.

such obstacles by exposing the disguised reality of capitalist society. Moreover, the rationality of such a choice depends upon the credibility of a future society delivering the promised benefits, and the possibility remains of a divergence between the interests of individuals and those of human-kind in general, given, for example, the short-term costs of minimising longer-term ecological problems. For these reasons, arguments for change (or revolutionary propaganda) may need to be both more concrete than Marx allowed in addressing the shape of a future society, and more value-oriented than he envisaged, promoting, for example, the idea to which Marx himself subscribed, that the well-being of individual humans depends upon their relation to other individuals and to humanity as a whole. The success of such interventions cannot be assumed, but their possibility makes the selection of new structures less improbable than Brenner suggests.

5.6 Conclusion

What I have argued in this chapter is that technological development can take a variety of forms with differing ecological consequences; that the role played by that development within historical materialism gives us no reason to suppose that adherents of the theory are committed to its damaging forms; and that it is possible to explain the occurrence of that development, upon which Marx's theory depends, in a way which allows for variations in its form, so that it may, depending on circumstances, include the kinds of development that will reduce the negative ecological consequences of productive activity. What links these various forms as *developments* of technology, and of the productive forces, is that they all in some way advance the capacity of humans to meet their needs or solve their problems. The variation occurs because the needs and problems of humans differ according to the social and material circumstances in which they find themselves, and it is because these circumstances can include the actual or potential occurrence of ecological problems that the capacity of technology to avoid or ameliorate such problems can form a part of the criterion for a Marxian conception of the development of the productive forces. This is not to say that a technological innovation that produces ecological problems can never be considered a development of the productive forces, but it does mean that, in assessing whether a particular innovation counts as a development, any ecological drawbacks need to be offset against its benefits.[87]

---

[87] The idea of offsetting ecological against other effects raises the problem of comparing dis-similar effects. This difficulty, however, is not confined to ecological issues, but occurs even with the traditional account of productive development as labour productivity. Labour

The account presented here is conducive to an ecological Marxism in that it both allows the avoidance or amelioration of ecological problems to be included among the criteria for the development of the productive forces, and at the same time suggests that the channelling of productive development in an ecologically advantageous direction may not be realisable at will but may depend on the selection of appropriate social structures, and that the range of technologically possible solutions actually available to a society may therefore be restricted by the choice of structures open to it. However, this variation in the potential for different kinds of productive development under different social structures takes us back to a reservation registered earlier in the argument. In discussing the Enabling Effect it was acknowledged that although there are reasons for interpreting modestly the expansion of productive output that Marx envisages as necessary for the development of socialism, this is a matter which warrants further investigation. The reasons for a modest interpretation of that expansion have to do with the satisfaction of needs, which Marx sees as its purpose. If our needs include such things as increased leisure time and a healthy and aesthetically satisfying environment, then we have reason to limit the expansion of output and to redirect technological development towards these ends. There is, however, at least one persistent element in Marx's discussion of needs which challenges this argument – his commitment to the *growth* of needs, which he appears to regard as both desirable and inevitable. Further investigation of this notion is therefore necessary before definitive conclusions can be drawn, and this will be undertaken in the next and final chapter. Also noted above was a problem arising from the suggestion that particular qualitative developments of the productive forces may be implicated in the Enabling Effect as conditions for the establishment of a communist society. The problem is whether it is possible for the productive forces to be developed in a way which simultaneously satisfies *all* of these conditions, for example whether technologies with reduced ecological impact can also be less alienating than present

productivity is defined by the ratio of quantity of product to the quantity of labour required to produce it. Comparison of different methods of producing the same product are straightforward, but when the types of product have changed it is not so simple. In these circumstances we may be able to consider products which satisfy the same needs, but the set of needs satisfied may also have changed. In some cases the comparison of ecological and other effects of a technology will be straightforward, for example when the ecological effects can be costed in terms of loss of production, or labour time required to rectify the ecological problem. In other cases, such as those involving aesthetic considerations or 'quality of life', comparison may be more difficult, but no more so than any other comparisons involving these sorts of needs. For a discussion of the problems of quantifying labour productivity, see Cohen 1978, pp. 56–9.

technologies. Since these qualitative preconditions are grounded in the idea of a communist society as one in which human needs are satisfied, we may expect that this problem too will be illuminated by the examination of Marx's conception of human need in the following chapter.

6 Capitalism, socialism and the satisfaction of needs

In the last chapter I examined the reasons for Marx's commitment to the development of the productive forces in order to ascertain whether his theory commits him to ecologically damaging forms of productive development, or whether, on the other hand, the effects that he attributes to the development of the productive forces can be achieved by ecologically benign forms of technological development. Two such effects were identified: the Undermining Effect whereby development of the productive forces undermines the viability of existing social forms, and the Enabling Effect whereby productive development makes possible the establishment of new social forms. The Undermining Effect, I argued, gives us no reason to attribute to Marx a commitment to ecologically damaging forms of productive development. The implications of the Enabling Effect, however, were less clear. Marx's view of communist society as a society of increased leisure provides a reason for thinking that forms of productive development designed to reduce labour time rather than forms designed to increase material production will be called for; but, in the *German Ideology* and particularly in the *Critique of the Gotha Programme*, Marx argues that an 'abundance' of material wealth will be necessary for the establishment of a communist society. It is worth quoting in full the relevant passage from the latter, which lies at the heart of the problem to be investigated in this chapter:

In a higher phase of communist society, after the enslaving subordination of the individual to the division of labour, and therewith also the antithesis between mental and physical labour, has vanished; after labour has become not only a means of life but life's prime want; after the productive forces have also increased with the all-round development of the individual, and all the springs of co-operative wealth flow more abundantly – only then can the narrow horizon of bourgeois right be crossed in its entirety and society inscribe on its banners: from each according to his ability, to each according to his needs![1]

---

[1] *Critique of the Gotha Programme*, pp. 320–1.

Three points should be noted here. First, abundant production is just one of the conditions identified by Marx as necessary for a communist society, alongside changes in the nature of work and in people's attitudes to it. Second, the abundance referred to is relative and not absolute; the springs of co-operative wealth, Marx says, must flow *more* abundantly. *How much* more abundantly they must flow is the question that must be considered here. The third point (which indicates how we may go about answering this question) is that Marx's reason for viewing greater abundance as necessary for the establishment of communism is that it is a necessary condition for the implementation of its principle of distribution according to needs. The degree of abundance required will therefore depend upon what needs it is that are to be satisfied under the communist distributive principle. That is what I will be investigating in this chapter.[2]

It might be objected that the principle of distribution according to needs does not commit Marx to the satisfaction of any particular need, but merely to the view that need should serve as a criterion for the allocation of distributive shares of whatever goods are available. Marx's point, however, is that this distributive principle can only be made to stick if everybody's needs can be satisfied at least to some minimum level. Without that prerequisite, as Marx says in the *German Ideology*, communism would merely result in generalised poverty, leading to renewed conflict over scarce resources.[3] So some level of abundance is necessary in order to reduce the intensity of competition as a condition for the implementation of the communist distributive principle. More than that, however, Marx does not view the principle 'To each according to his needs!' merely as a matter of distributive shares. As we shall see, he is scornful of a 'crude communism'

---

[2] Contrary to the approach followed here, Philippe Van Parijs (1993, pp. 211–32) claims that communism requires *absolute* abundance (defined as the capacity of an economy to satiate simultaneously its members' appetites for consumption and leisure, and dismissed by Van Parijs as hopelessly utopian). This claim, however, derives from Van Parijs's contentious interpretation of the communist slogan as implying that 'people provide their labour spontaneously, for no pay, while all their material wants are satisfied, thanks to the free provision of all the goods they care for' (p. 214). This departs from Marx's formulation in taking communism to be a society in which not just needs but wants are provided for unconditionally. Moreover, while Van Parijs allows that communism will provide 'guidance' to its members – e.g. guiding them to do the most amount of work that they don't mind doing for free – he holds that such guidance can only relate to choices between which the individual is indifferent; 'no self-sacrifice is required of the communist (wo)man' (p. 218), since this would conflict with the unconditional satisfaction of wants, including the desire for leisure. Some sacrifice of desired leisure is, however, consistent with the communist slogan if it is interpreted as promising only as much leisure as is *needed*; indeed, notwithstanding the reference to labour becoming life's prime want, the possibility of being required to work more than one might desire appears to be implicit in the first part of the communist slogan: 'from each according to his ability'.

[3] *The German Ideology*, p. 56 (see text to note 64, ch. 5 above).

which merely advocates that sparse goods be fairly redistributed. What Marx wants to see is a society in which the needs of everybody can be satisfied to a degree that is truly worthy of human beings.

It is clear, just from the quoted passage, that Marx anticipates some increase in material production as a prerequisite for the establishment of a fully developed communist society. This is not sufficient reason, however, to condemn his project as ecologically unsustainable. For one thing, the increase that Marx anticipates is relative to nineteenth-century levels of production, and it is open to question whether a further increase from present-day levels of production is necessitated by Marx's theory of communism. Moreover, even if Marx's theory does commit him to further expansion of material production, this will not necessarily lead to increased environmental damage. As was shown in the last chapter, advances in technology may produce efficiency gains, allowing more to be produced with an ecological impact which is no worse and possibly better than before. So, if Marx is committed to an increase in output then the ecological sustainability of his programme will depend upon the extent of that increase, and this in turn will depend upon the extent of the needs that the increased production is to serve. It should be noted, however, that the relation between need-satisfaction and material production is far from straightforward. On the one hand the existence of unmet needs would appear to demand an increase in material production with the potential that this brings for increased environmental damage. But on the other hand, as has been noted earlier, the preservation or restoration of a healthy environment will also count as the satisfaction of human need. And so, perhaps, will the reduction of working hours which Marx advocates, and the changed character of labour called for in the passage quoted above. In order to assess the ecological implications and viability of Marx's theory it is therefore necessary to consider the role of needs in its qualitative as well as its quantitative aspects.

I will begin by outlining the structure of the concept of need. This will serve as a framework for the subsequent investigation, which will comprise, firstly and most extensively, an analysis and ecological appraisal of Marx's most explicit and detailed account of human needs, that given in his *Economic and Philosophical Manuscripts*, and, secondly, a briefer examination of some aspects of the concept as it appears in Marx's later works.

6.1 The concept of need

It has become almost a commonplace in discussions of need to observe that needs-claims have the triadic structure '*A* needs *X* for (or in order to)

Y'.[4] What this signifies is that a thing that is needed (a satisfier, X) is always needed as a condition for obtaining some end or good. Statements of the form 'A needs X', it is argued, are elliptical for 'A needs X for Y'. This structure gives us a way of distinguishing different classes of needs: needs can be classified according to the kinds of end that they serve. For example: there are things that we need in order to survive, such as air, water and food. There are things that are needed for a healthy existence: *clean* air and water, *nutritious* food, etc. There are things that we need in order to function, or to thrive, in a modern society: a home, means of transport, education, a job, or whatever. And there are things that are necessary for the fulfilment of our personal projects and desires. There may also be things that we need in order to satisfy universal ends, which we share as human beings. Generally, we can distinguish between the *concept* of need given by the triadic structure itself, the various *conceptions* of need adhered to by different theorists, and different *classes* of needs included within any particular conception. Different theorists' conceptions of need (including Marx's) can then be characterised in terms of the classes of need (and correspondingly the range of ends) that they recognise.

A further implication of the triadic understanding of the concept of need concerns the relation between needs and wants. These are often regarded as being on a continuum, but according to the triadic account they form two distinct categories. While all needs have the form 'A needs X for Y', only some wants have the parallel form 'A wants X for Y', since wants may be divided into those things that are desired as means to other ends (i.e. instrumentally) and those things that are desired as ends in themselves (i.e. intrinsically), and the structure 'A wants X for Y' applies only to the former. Where something is desired as an end in itself the structure is simply (and non-elliptically) 'A wants X'. Moreover, even our instrumental wants – which do have a triadic structure – can be sharply distinguished from needs by attending to the way in which this structure applies to the two relations. To want something is to be in an intentional mental state whereby one is motivated to try to get that thing, whereas to need something is for it to be objectively the case that that thing is necessary for the achievement of some end or good. A's *wanting* X for Y thus depends on A's beliefs about X, and in particular her belief that it is a means to the achievement of Y. A's *needing* X, on the other hand, depends not on any of A's *beliefs* about X, but on the *fact* that X is necessary in order to achieve Y. A consequence of this is that need-statements and want-statements, despite

---

[4] See, for example, Doyal and Gough 1991, p. 39; Barry 1965, pp. 48–9; Plant *et al.* 1980, p. 26.

their shared triadic structure, differ formally in their referential opacity. In a need-statement we can substitute any extensionally equivalent term for X without altering the truth-value of the statement, but in a want-statement we cannot. As David Wiggins puts it: 'If I want to have x and $x = y$, then I do not necessarily want to have y. If I want to eat that oyster, and that oyster is the oyster that will consign me to oblivion, it doesn't follow that I want to eat the oyster that will consign me to oblivion. But with needs it is different. I can only need to have x if anything identical with x is something that I need.'[5]

Needing thus is a referentially transparent context while wanting is referentially opaque.

The importance of this in the present context is that it shows wanting and needing to be distinct concepts with distinct truth conditions. Even where the wants and needs in question are directed towards the same ends, it is possible to want things that are not needed or to need things that are not wanted. For example, given that I have a desire to be fashionable (either an intrinsic or an instrumental desire), it may be the case that I *want* a beret in order to be fashionable, but that what I *need* in order to be fashionable is a baseball cap, which I don't want. We can, therefore, acknowledge that Marx endorses some notion of need-satisfaction as a normative principle, and set out to investigate that notion on the basis of the triadic concept, without supposing – as some of his green critics appear to suppose[6] – that he is thereby committed to the view that people should be provided with whatever they want.

But neither – quite rightly – does the triadic account rule out this possibility. Not only does the triadic account allow the obvious fact that our needs and wants often do coincide; it also enables us – by including the satisfaction of desires among the relevant ends (Y) – to specify a conception of needs which by its definition includes some or all of the things that we want. We can, therefore, adopt the triadic concept of need as a framework for the investigation of Marx's use of this concept, without prejudging the suggestion that Marx uses it in such a way as to include the satisfaction of wants among our needs.

The triadic account of need appears then to provide a useful framework for the investigation of Marx's use of this concept. There are, however, two suggestions to be found in the literature which appear to challenge its validity for this purpose, by showing that the triadic structure is not

---

[5] Wiggins 1987, p. 6. See also Doyal and Gough 1991, pp. 41–2.
[6] For example, Irvine and Ponton (1988, p. 80) refer to the principle 'to each according to his needs' as a 'blank cheque'.

applicable to all needs. First, it is suggested that there is a sense in which 'need' refers to a 'drive' or 'motivational force',[7] a well-known example being Maslow's hierarchy of needs. Offered as a theory of motivation, Maslow's hierarchy consists of five categories of 'basic needs' – physiological needs, safety needs, belongingness and love needs, esteem needs, and the need for self-actualisation – and the claim that these needs are 'organized into a hierarchy of relative prepotency' such that the first set of needs dominates the behaviour of the individual until such a time as they are satisfied, whereupon the second set of needs emerges and comes to dominate, and so on.[8] Powerful criticisms have been levelled against Maslow's theory,[9] but my purpose here is not to establish whether the theory is true, but whether Maslow's use of the term 'need' undermines my proposal to base my investigation of Marx's concept of need upon the triadic structure. There are in fact two ways in which the idea of needs as drives may be interpreted, and neither, I suggest, undermines this proposal.

First, we can understand the idea of needs as drives as the claim that the term 'needs' sometimes refers to ultimate, non-instrumental motivations. Paul Taylor, for example, states that 'need' is used 'to describe motivations, conscious or unconscious, in the sense of wants, drives, desires, and so on'.[10] It is true that needs/drives understood as non-instrumental motivations will not fit into the triadic structure, since this structure specifically expresses needs as relations of instrumentality. It seems clear, however, that if we do use 'need' in this way we use it in a sense that is distinct from that discussed above, a sense which has more in common with 'wants' than with our normal uses of 'needs'. This can be seen in the fact that, just as one can want something one doesn't need or need something one doesn't want, so also one can be 'driven' to consume things that one does not need (like lots of alcohol) and can need things (like exercise or a healthy diet) for which one has no 'drive'.[11] The intuitiveness of this formulation also suggests that this sense of 'need' (if it is a literal sense at all) is a peripheral one. What matters, however, is that the sense which Taylor attributes to 'need' is distinct from the sense discussed above and that it is not a sense that can plausibly be read into Marx's use of the communist slogan. It will become clear (in the following section) that however Marx did intend 'To each according to his needs!' to be understood, he did not intend it as an endorsement of whatever urges or 'drives' a person happens to have.

---

[7] Doyal and Gough 1991, pp. 35–9.
[8] Maslow 1970, especially ch. 4. Theory of motivation: pp. xi, 35; 'relative prepotency': p. 38.
[9] For criticisms of Maslow, see Doyal and Gough 1991, p. 36.
[10] Quoted in Springborg 1981, p. 254.
[11] Cf. G. Thompson, quoted in Doyal and Gough 1991, p. 36.

The second way of understanding the idea of needs as drives is suggested by a comment of Maslow in which he distinguishes (albeit not very
clearly) between needs and their goals.[12] What this suggests is not that
'need' refers to a person's ultimate motivation or goal, but that we need
certain things *in order to achieve our goals or to satisfy our drives*. This interpretation will, like the previous one, prove to be implausible as an interpretation of Marx, but this account in any case fits into the triadic structure
and therefore presents no threat to the proposed method of investigation.
On this account, needs are relations of the form '*A* needs *X* for *Y*', where *Y*
is the satisfaction of some drive such as those postulated by Maslow's
psychological theory.

A more interesting objection to the use of the triadic framework for the
investigation of Marx's conception of needs arises from the normative use
to which Marx puts the concept. It has been argued that triadic needs,
being merely instrumental relations, lack normative force. This has been
used to rebut claims that the concept of need can serve as a basis for normative discourse capable of bridging the 'fact–value gap'. Against such
claims it is argued that any normative force that statements of need appear
to have derives not from the need statement itself but from a further claim,
a hidden evaluative premise, concerning the desirability of the end to
which the need is directed. We may agree, in other words, that *A* needs *X*
for *Y*, but whether this provides a reason for *A* to be given *X* will depend
entirely upon the evaluation of *Y*.[13] The fact that Marx does regard needs
as providing a normative basis for his project, without attempting further
justification of any ends served by those needs, creates doubt about
whether in fact the sense in which Marx uses 'need' is the same as the sense
which is captured by the triadic analysis. It could of course be that Marx
does use 'need' in the sense given by the triadic analysis but that he is con-
fused about its normative status, or that he holds the ends implicit in his
need-claims to be self-evidently valuable and therefore not requiring justification. A more plausible account, however, may be derived from the
analysis of need-claims proposed by David Wiggins.

Wiggins agrees that there is a class of need sentences which have the
instrumental structure '*A* needs *X* for *Y*' and another class which have the
same logical structure but are expressed elliptically in the form '*A* needs
X'. For example:

---

[12] Maslow 1970, p. 38. At other times Maslow clearly fails to make this distinction, e.g. p. 35:
    'The needs that are usually taken as the starting point for motivation theory are the so-
    called physiological drives.'

[13] See, for example, Barry 1965, pp. 48–9. Springborg 1981, pp. 252–74 surveys a number of
    recent views on the nature and normative status of needs.

Someone may say 'I now need to have £200 to buy a suit', or, speaking elliptically, 'I need £200'. If he can't get the suit he has in mind for less than £200, then it is true, on an instrumental reading of his claim, that he needs £200.[14]

Wiggins argues, however, that there is another class of need sentences which have the form 'A needs X', but which are *not* elliptical for 'A needs X for Y'. This use of the word is illustrated by a possible reply to the man who claims to need £200 in order to buy a suit:

We can properly and pointedly respond to his claim with: 'You need £200 to buy that suit, but you don't need £200 – because you don't need to buy that suit'.[15]

What we are denying here, Wiggins suggests, is not that there is some further end for which the suit is necessary (if the man were to respond to us by identifying some such end it would show that he had misunderstood our objection); what we are denying is 'that he *cannot get on without that suit*, that *his life will be blighted without it*, or some such thing'.[16] This is what Wiggins calls the *absolute* or *categorical* sense of needing, and he argues that it is when 'need' is used in this sense that it has the practical authority associated with the term.

Now, it might seem that Wiggins has rescued the normative force of the term 'need', but at the cost of forcing us to abandon the triadic structure as a tool for investigating Marx's use of the concept. This, however, would be to misunderstand Wiggins's point, which is not that 'need' in the absolute sense is unconcerned with ends or purposes, but that when 'need' is used in this sense the purpose is internal to, and therefore fixed by, the meaning of the word. What it means to say that something is needed absolutely is that it is something that must be had in order to avoid harm; thus, as Wiggins puts it:

I need [absolutely] to have x

is equivalent to:

I need [instrumentally] to have x if I am to avoid being harmed.

The triadic formula 'A needs [instrumentally] X for Y' thus serves to express the internal structure of the concept of absolute needing, where (in virtue of the meaning of absolute needing) Y is the avoidance of harm. And it will be as well to keep this structure in mind even once it has been decided that, in a particular case, 'need' is being used in its absolute sense, since the idea of *harm* is itself problematic and in need of further elucidation. We might say that the absolute sense of 'need' fixes Y as the avoid-

[14] Wiggins 1987, p. 7.   [15] *Ibid.*, p. 8.   [16] *Ibid.*, p. 9.

ance of harm, but that this is not sufficient to fix it absolutely: the interpretation of harm, and therefore the interpretation of [absolute] need, will depend upon one's ideas about what it is for a person to flourish or for a person's life to be blighted.[17] So, whether a particular use of 'need' is instrumental or absolute, we may hope to elucidate its content by considering the end (Y) to which that need relates.

Marx's main uses of 'need', it will become apparent, do relate to the loosely defined ends of human flourishing and the avoidance of harm which Wiggins associates with the categorical sense of the term. But, since it is the content of Marx's need statements that we are concerned with – the range of things he thinks people need – rather than the meaning of the word, the question of whether Marx uses the term in an instrumental or an absolute sense will be secondary to the identification in more specific terms of the relevant ends.

6.2 Marx on true and false needs

In the previous section I claimed that Marx's use of the slogan 'To each according to his needs!' was not intended as an endorsement of whatever urges or 'drives' a person happens to have. This view is supported by passages from several of Marx's major works, in which it appears that he distinguishes between 'true' and 'false' needs[18] – that is, between things that people really do need and things that they falsely believe themselves to need – and criticises capitalism for its tendency to induce false needs. In this Marx appears to be in agreement with the views of the contemporary green theorists summarised by Andrew Dobson: 'they make an (unoriginal) distinction between needs and wants, suggesting that many of the items we consume and that we consider to be needs are in fact wants that have been "converted" into needs at the behest of powerful persuasive forces. In this sense they suggest that little would be lost by possessing fewer objects.'[19] The underlying thought here is that a commitment to the satisfaction of whatever needs people perceive themselves to have would be ecologically unsustainable in view of the apparently unlimited expansion to which such perceived needs are subject. A distinction between actual and perceived needs would therefore appear to be necessary if the satisfaction of needs is to be an ecologically sustainable objective. It appears, therefore, that in distinguishing between true and false needs

[17] Cf. *ibid.*, p. 11.   [18] These are not Marx's terms, as we shall see.
[19] Dobson 1990, p. 18.

Marx is making the first step towards an ecologically sustainable interpretation of the communist slogan 'To each according to his needs!' Before exploring this suggestion, however, we need to substantiate the claim that Marx's use of the slogan is restricted to 'true' needs, since this interpretation has been rejected, both by Marxist theorists (of which more below) and by green critics who view the communist slogan as a 'blank cheque' offering unlimited consumption.[20]

Marx's fullest account of human needs is found in the *Economic and Philosophical Manuscripts*. In a series of passages in this work Marx not only acknowledges but insists upon a distinction close to that proposed by the greens. For example, he writes that under capitalism, 'every person speculates on creating a *new* need in another, so as to drive him to fresh sacrifice, to place him in a new dependence and to seduce him into a new mode of *enjoyment* and therefore economic ruin. Each tries to establish over the other an *alien* power, so as thereby to find satisfaction of his own selfish need.'[21] Of course, Marx does in this passage use the term 'need' to refer to the appetites so created, but it is clear that these are to be distinguished from the needs whose satisfaction he advocates through his use of the communist slogan. Moreover, the description of such appetites as 'imaginary', and as 'fantasy, caprice and whim', 'depraved fancies', etc., in the following passage suggests that Marx does not view these appetites as real needs at all, or at least not in the sense in which he primarily uses the term:

the extension of products and needs becomes a *contriving* and ever-*calculating* subservience to inhuman, sophisticated, unnatural and *imaginary* appetites. Private property does not know how to change crude need into *human* need. Its *idealism* is *fantasy, caprice* and *whim*; and no eunuch flatters his despot more basely or uses more despicable means to stimulate his dulled capacity for pleasure in order to sneak a favour for himself than does the industrial eunuch – the producer – in order to sneak for himself a few pieces of silver, in order to charm the golden birds out of the pockets of his dearly beloved neighbours in Christ. He puts himself at the service of the other's most depraved fancies, plays the pimp between him and his need, excites in him morbid appetites, lies in wait for each of his weaknesses – all so that he can then demand the cash for this service of love. (Every product is a bait with which to seduce away the other's very being, his money; every real and possible need is a weakness which will lead the fly to the glue-pot. General exploitation of communal human nature, just as every imperfection in man, is a bond with heaven – an avenue giving the priest access to his heart; every need is an opportunity to approach one's neighbour under the guise of the utmost amiability and to say to him: Dear friend, I give you what you need, but you know the *conditio sine qua non*; you know the ink in which you have to sign yourself over to me; in providing for your pleasure, I fleece you.)[22]

---

[20] Irvine and Ponton 1988, p. 80.   [21] *Economic and Philosophical Manuscripts*, p. 101.
[22] *Ibid.*, p. 102. It may appear from this passage, and from what Marx says elsewhere in the

Although the *Manuscripts* contain Marx's most extensive account of human needs, it is not only in this work that Marx distinguishes between true and false needs. A central theme in *Capital*, for example, is the distinction between production aimed at the creation of use-values and production aimed at the creation of value or (what amounts to the same thing) at the expansion of capital. The latter, Marx asserts, has a tendency to expand independently of human needs.[23] Indeed, Marx characterises this tendency as a 'need' of capital to which human needs are subordinated, capitalism being 'a mode of production in which the worker exists to satisfy *the need of the existing values for valorization*, as opposed to the inverse situation, in which objective wealth is there to satisfy the worker's own need for development'.[24] But capital's 'need' to expand is not only independent of human needs; it moulds the needs that human agents perceive themselves to have. Owners of capital need to expand production if they are not to be ejected under pressure of competition from the class of

Manuscripts about the suppression of all but the most basic needs of the workers, that he recognises the development of false needs only among the non-working classes: one capitalist seeking to 'fleece' another, his friend and neighbour in Christ, to seduce away the money that is his very being, and to drive him to economic ruin. The following passage also appears to support this interpretation: 'the sophistication of needs and of the means [of their satisfaction] on the one side produces a bestial barbarisation, a complete, crude, abstract simplicity of need, on the other . . . Even the need for fresh air ceases to be a need for the worker. Man returns to a cave-dwelling, which is now, however, contaminated with the pestilential breath of civilisation' (*ibid.*, pp. 102–3).

On this interpretation we might be inclined to judge Marx's account outdated and to accept the suggestion of more recent writers in the tradition such as Marcuse, and of green critics of capitalism, that the phenomenon of false need touches all classes of contemporary capitalist society. Further comments by Marx, however, suggest an interpretation of the above passages which is consistent with the contemporary judgement; an interpretation in which workers' needs are not *simply* suppressed by the private property system since they too experience false needs, but *crude* false needs as opposed to the *sophisticated* or *refined* false needs experienced by their employers: 'Industry speculates on the refinement of needs, it speculates however just as much on their *crudeness*, but on their artificially produced crudeness, whose true enjoyment, therefore, is *self-stupefaction* – this *illusory* satisfaction of need – this civilisation contained *within* the crude barbarism of need. The English gin shops are therefore the *symbolical* representations of private property. Their *luxury* reveals the true relation of industrial luxury and wealth to man. They are therefore rightly the only Sunday pleasures of the people which the English police treats at least mildly.' (*Ibid.*, p. 107.)

[23] Marx compares his own distinction between use-value rationality and exchange-value rationality with Aristotle's distinction between *economics* as 'the art of procuring use-values for the household' and *chrematistics* as 'the art of money-making' (considered as an end in itself). The central point which Marx wishes to take from Aristotle is that whereas economics (use-value production) has a limit, 'riches, such as chrematistics strives for, are unlimited.' (*Capital*, vol. I, p. 253n.) This comparison with Aristotle is discussed in Collier 1994.

See also *Theories of Surplus Value*, part I, p. 282: The capitalist 'produces for the sake of production, he wants to accumulate wealth for the sake of the accumulation of wealth. In so far as he is a mere functionary of capital, that is, an agent of capitalist production, what matters to him is exchange-value and the increase of exchange-value, not use-value and its increase.'   [24] *Capital*, vol. I, p. 772; my emphasis.

capital-owners, and it is in order to satisfy this 'need' that they in turn seek to create false needs in others.[25] The indifference of capitalist production to *genuine* human needs is expressed by Marx in the following definition of its basic element, the commodity: 'The commodity is, first of all, an external object, a thing which through its qualities satisfies human needs of whatever kind. The nature of those needs, whether they arise, for example from the stomach, *or the imagination*, makes no difference.'[26] Again, we may suppose that needs arising from the imagination are not real needs at all, on Marx's account, but imagined or false needs.

In ascribing to Marx a distinction between true and false needs I am not denying that the term 'need' has a legitimate and literal application to the phenomena that I take Marx to be identifying as false needs. If I am an inhabitant of a society where producers speculate on seducing me into new modes of enjoyment, it may be that possession of the latest product really is instrumentally necessary for the satisfaction of my desires, or for me to avoid feelings of frustration and deprivation, or for the maintenance of my status in the community (keeping up with the Joneses). And similarly the producer really does need to expand his capital as a condition of remaining a capitalist. These, according to the triadic analysis above, are legitimate uses of the term 'need'; it is just that the ends implicit in these instrumental need-claims are not the ends that Marx holds to be constitutive of human flourishing and which for him define the sense (or, better, the conception) in which need serves as a normative principle in the communist slogan.

One writer who disputes the interpretation presented here, according to which Marx presents a distinction between true and false needs, is Agnes Heller. She sees that distinction, both in theory and in the practice of (then) existing socialist societies, as a basis for dictatorship:

---

[25] A difference between the 'needs' that capital directly induces in its owners and those that they in turn induce in others, is that the former are genuine instrumental needs, necessary for the maintenance of the capitalist's economic position, if not for his human flourishing. The latter, by contrast, may be pure illusion.

[26] *Capital*, vol. I, p. 125; my emphasis. Patricia Springborg (1981, p. 106) reads the second sentence of this passage as evidence that in *Capital* Marx was no longer concerned to press the distinction between needs and wants, a reading which may appear to be supported by the quotation from Barbon that Marx appends to his definition: 'Desire implies want; it is the appetite of the mind, as natural as hunger to the body.' However, an alternative reading, more in tune with Marx's assertions elsewhere of a distinction between true and false needs, is that Marx is here expressing not his own view but that which he finds implicit in the workings of capitalism, and therefore reflected in the writings of political economists like Barbon. What Marx is expressing, on this reading, is the irrelevance to capitalist production of the distinction between true and false needs, since it makes no difference to the determination of value in the capitalist economy whether or not the need served by a commodity is genuine. And since, as we have seen, Marx criticises capitalism for this indifference we may suppose that he does regard the distinction as important.

The mere gesture of separating 'real' needs from 'imaginary' ones forces the theoretician into the position of a god passing judgement upon the system of needs of society. One can separate real needs from the imaginary only by assuming that one *knows* which are the 'real', the 'true' ones. When the non-reality of needs is explained by the theory of manipulation, the knowledge of the theoretician passing judgement can only originate from the fact that his consciousness is not fetishized, that it is 'the' correct consciousness. But how does the theoretician know that his consciousness is 'the' correct one?[27]

The danger of a 'dictatorship over needs' should not be ignored, but two points must be made about Heller's argument. The first is that in arguing against the true/false need distinction she runs together at least three distinct questions: whether, in theory, we can distinguish between true and false needs (whether, in other words, people can be mistaken about their needs); whether true needs as opposed to perceived needs should be used as a basis for the distribution of goods; and, if so, how such needs should be assessed, and by whom. It would be quite possible, for example, to decide that though individuals can be wrong about their own needs, the difficulty of ascertaining their true needs for them and the danger of dictatorship are sufficient reason to refrain from using true need as a distributive principle. Heller, however, goes beyond this to the extent of denying – absurdly – that people can be mistaken about their needs. She writes, for example, that 'What individuals are aware of to be their need, actually is their need. It is real, it has to be acknowledged, it has to be satisfied.'[28]

The second point is that, even were it true that the distinction between true and false needs is conducive to dictatorship, this would not be sufficient reason to refuse to acknowledge such a distinction in Marx. In order to recruit Marx as an opponent of that distinction some textual evidence is needed, and in the light of the passages quoted above the prospects do not look promising. Heller, however, accounts for such passages by claiming that what is depicted in them is not *false* need but *manipulated* need.[29] It may appear that manipulated needs are simply false needs under another name, but what is distinctive about Heller's account is her refusal to attribute to Marx any evaluative distinction among perceived needs under socialism. The distinction that she does acknowledge is between the manipulated and alienated needs felt under capitalism and the non-manipulated and unalienated needs that (she thinks) will be felt under socialism. The refusal to acknowledge a distinction between true and false needs under socialism may remove the danger of a dictatorship over needs, but it also deprives Marx of the ability to respond to ecological scarcity (or indeed any form of resource scarcity) by limiting the range of

[27] Heller 1985, pp. 285–6.   [28] *Ibid.*, p. 291.   [29] Heller 1976, pp. 50–1.

appetites to be satisfied. A supporter of Heller might argue that Marx has no need to impose such a limit, since people's perceived needs taken as a whole will be more modest once the manipulative influence of capitalism is removed. This, however, will not do: firstly because the occurrence and extent of this development is in itself very uncertain (Heller herself thinks that material needs will continue to grow under socialism, albeit more slowly than under capitalism); and secondly because, whatever the extent of such a reduction, it remains a contingent matter whether this will be sufficient to render the remaining appetites satisfiable within the limits of an uncertain level of ecological scarcity.

Heller's rejection of a distinction between true and false needs is therefore ecologically problematic. But it is also implausible as an interpretation of Marx. Heller appears to be rejecting an evaluative distinction (between true and false needs) in favour of a historical one (between the needs that are felt under capitalism and those that will be felt under socialism). There must, however, be an evaluative aspect to this distinction, since it is clear in the passages quoted above that Marx regards the existence of what Heller characterises as manipulated needs as one of the reasons for criticising capitalism.[30] We may therefore ask what it is about the 'manipulated needs' characteristic of capitalism that Marx objects to, and what reason there is to suppose that the features to which Marx objects cannot ever be possessed by the perceived needs that may develop under socialism.

It will not do to reply that 'manipulated needs' are *alienated*, and then to define alienation in terms of capitalist labour relations, or private property, or whatever; for it remains to be said what is wrong with alienation, and until that is determined we cannot rule out the possibility that some of the needs that develop under socialism will share the characteristics that make alienated needs undesirable. It cannot be simply the artificiality of 'manipulated needs' that Marx objects to, since many human needs, including those that Marx anticipates developing under socialism, are mediated by human activity. Indeed, as Heller herself realises, Marx does not regard the needs created by capitalism (and therefore artificial) as by that token automatically impugned, but 'emphasises the fact that capitalism creates needs that are "rich and many-sided" at the same time as it impoverishes men and makes the worker a person "without needs"', welcoming the former while deploring the latter.[31] It is not, then, the artificiality of 'manipulated

---

[30] The term 'manipulated needs' appears in quotation marks in the following assessment of Heller's account, in order to indicate my reservations about the appropriateness of this term as a description of the phenomenon discussed by Marx in the passages quoted above. The nature of my reservations and my reasons for preferring to retain the term 'false needs' will emerge from the criticism of Heller below.   [31] Heller 1976, p. 47.

needs' that for Marx disqualifies them from being worthy of satisfaction under the communist slogan. Neither is it plausible to suppose (as Heller's later work suggests) that Marx's objection to 'manipulated needs' is based on the Kantian thought that, in having their needs manipulated, people are treated as mere means for the satisfaction of others' ends. For though the intention to use others as means surely would have been objectionable to Marx, it is Marx's view that the capitalists who direct and benefit from this manipulation very often do so not from any intention to subordinate others' interests to their own, but in the mistaken belief, induced by their own ideology, that their interests and those of the rest of society are the same. Marx's objection to 'manipulated needs' therefore cannot be solely an objection to such intentions; he must object also to the reality of such subordination whether intentionally produced or not.

In fact, Heller herself indicates what it is about 'manipulated needs' that Marx objects to, but in doing so undermines her claim to have eliminated the notion of true and false needs from Marx's theory. She introduces the idea of 'manipulated needs' by suggesting that, as the development of needs becomes, under capitalism, a means of expanding capital and furthering the interests of its owners, so there arises the possibility that people will be manipulated into feeling a need for things which are incompatible with the development of the 'rich and many-sided' human needs to which capitalism also gives rise.[32] But what we have here are two possible perceptions of needs, either of which may be realised but not both, and it only makes sense for Marx to reject one set of perceived needs on the grounds that it inhibits the development of the other if we suppose that he regards the latter as more worthy of development than the former. More specifically, his objection to the situation in which the presence of 'manipulated needs' inhibits the development of 'rich and many-sided needs' depends upon the belief that a failure to develop and satisfy rich and many-sided needs would be a loss to the individuals concerned, whereas a failure to develop and satisfy those 'manipulated needs' that are incompatible with rich and many-sided needs would not. Once we say this we are not far from saying that, for Marx, rich and many-sided needs are *real* needs independently of whether they are perceived as such, while the others are merely *perceived* as needs, with no such objective status.

Though Heller fails to eliminate the distinction between true and false needs from Marx's theory she does bring to light an important difference between Marx's account of the distinction and that of the green theorists

---

[32] *Ibid.*, pp. 50–1.

surveyed by Dobson. Marx shares with the greens the view that true needs should be distinguished from those things that agents believe or feel themselves to need, but (as Marx's reference to 'rich and many-sided' needs suggests) he is more inclined to think of true needs as extensive and expanding, while they are more inclined to think of them as modest and constant. The difference should not be overstated; the truth is likely to be that true needs are in some respects narrower and in other respects wider than our perceived needs, and this is something that both parties recognise.[33] Furthermore, while it is true that in general Marx places more emphasis on the latter aspect and the greens on the former, it should be remembered that (as outlined in the introduction to this chapter) an increase in need, even if it means an increase in material production need not be ecologically damaging. This difference of emphasis between Marx and the greens does, however, indicate the need to develop a positive account of what true need might amount to for Marx. This task will be pursued in the following sections.

6.3 Animal needs, workers' needs and human needs in the *Economic and Philosophical Manuscripts*

We have seen in the previous section that Marx is prepared to distinguish between the 'needs' that people feel themselves to have and the needs that they actually have. Some of the 'needs' that people feel, Marx believes, are not genuine needs and therefore lack the normative force of genuine needs. Some of the 'needs' induced by capitalism actually inhibit the development and satisfaction of a 'rich and many-sided' set of truly human needs.

Apart from the epithet 'rich and many-sided', Marx's conception of truly human needs has so far been characterised negatively, by reference to the contrast with false needs. In this section and the next, I will develop a positive account of human needs, based on the *Economic and Philosophical Manuscripts* which, as noted above, contain Marx's most extensive reflections on the subject. I will for the moment leave aside questions concerning the continuity between the account of needs in the *Manuscripts* and the account implicit in Marx's later writings, my intention here being to show that there is available to Marx an account of human need which is compatible with the ecological interpretation of historical materialism developed in the previous chapter. Whether this account is compatible with the more fragmentary references to human needs which appear in Marx's later

---

[33] See, for example, Porritt 1985, pp. 197–8.

works, and whether these works call for a revision or reconsideration of the account developed here, are questions that will be considered subsequently.

The *Economic and Philosophical Manuscripts* are best known for the theory of alienation that they contain. Central to that theory is an account of human nature and human needs which emphasises the ways in which the nature and needs of humans differ from those of (other) animals. But just as commentators have been divided over the significance of this 'humanist' perspective ever since the *Manuscripts* were first published in 1932, so more recently have the *Manuscripts* prompted opposing responses amongst ecologically minded commentators. Some have seized upon the *Manuscripts* as evidence of an 'ecological Marx', one who criticises capitalism for the 'alienation from nature' that it engenders and who presents a vision of communism as 'the genuine resolution of the conflict between man and nature and between man and man'. Others, however, have objected to the emphasis that Marx places upon the differences between humans and animals, and to his account of the 'mastery', 'transformation' or 'humanisation' of nature by humans.[34]

One of the recurring themes of this book has been that notions such as the 'mastery' or 'transformation' of nature (along with that of the development of the productive forces) need not be ecologically problematic if they are interpreted in relation to human interests and needs, and if it is recognised that these include a need for and interest in a habitable and flourishing environment. My purpose here is to discover whether the conception of human needs that emerges from the *Manuscripts* does give adequate regard to their ecological component. First, however, I will attempt to pre-empt objections arising from the animal/human contrast which Marx uses to frame this conception of needs.

One writer who acknowledges both sides of the argument about the ecological standing of the *Manuscripts* is Ted Benton. On the one hand, Benton notes with approval Marx's view of non-human nature not just as a means to satisfy bodily needs but as 'a source of spiritual and aesthetic nourishment', and his view of communism as a society in which this dimension of the human–nature relation can flourish.[35] On the other hand, he sees Marx's writings on this subject as 'deeply ambivalent and often contradictory'. Some of Benton's worries on this score have been discussed in previous chapters, but among them is Marx's aforementioned view of the

---

[34] See, for example, the exchange between D. C. Lee, Val Routley and Charles Tolman in *Environmental Ethics*, 1980–82.

[35] Benton 1992, p. 68; cf. Benton 1989, pp. 53–5 and Benton 1993, p. 24.

uniqueness of human needs, a view which Benton rejects as a form of 'speciesism'. In opposition to this view, Benton recommends that the specific features of humans should be differentiated and elaborated 'from an initial recognition of the common core of "natural being" which we share with other living creatures', and that, in order not to revert to a human–animal 'dualism', we should recognise that 'those things which only humans can do are generally to be understood as rooted in the specifically human ways of doing things which other animals also do'.[36]

This formulation appears to me to contain an ambiguity, located in the expression 'rooted in'. If it is intended as a causal statement about the historical process whereby humans develop their unique needs and powers, then it seems both unexceptionable and consistent with Marx's view that humans undergo a historical process because of the particular ways in which they, as conscious, intelligent and social beings, go about meeting their needs. However, what Benton appears to mean is not that human needs arise *as a result of* animal needs being met in a peculiarly human way, but that they simply *are* animal needs met in a human way.[37] This, it seems to me, is an excessive claim. Benton's concern at this point, it must be said, is not so much the ecological implications of Marx's humanism as its implications for our moral responsibilities in the treatment of animals, hence his use of the term 'speciesism'. But while he is surely right to argue that the morally relevant similarities between us and other animals are often underestimated, it appears to be a strained and unhelpful use of language to insist (*a priori*?) that every human need be subsumed under some category of animal need. In any case, what matters in assessing the ecological sustainability of Marx's conception of human need is not the fact that for Marx human needs differ from animal needs, nor even the extent to which they differ, but the *content* of the human needs that emerge from this comparison. A distinction is required here between description and determination. Marx *describes* the content of human needs by comparison with animal needs, but the relation is an external and contingent one; human needs are not *determined* (either logically or causally) by the animal needs with which they are contrasted. Thus, a change to Marx's account of animal needs, giving greater recognition to the commonalities between animals and humans, might require a change in the way human needs are

---

[36] Benton 1993, pp. 47, 48.

[37] This interpretation is suggested by Benton's questioning of what he takes to be Marx's view that human needs 'constitute a separate, *sui generis* class of needs, set over and above our "animal" needs, and peculiar to us as humans' (Benton 1993, p. 48).

described, but it would leave the content of those needs, and therefore their ecological consequences, unchanged.[38]

At the core of Marx's account of human needs in the *Manuscripts* is the claim that genuine human needs exceed what is necessary for bodily survival. Capitalism, Marx argues, denies this. In order to elucidate this claim we may use the term 'socially acknowledged needs' to refer to those needs that are acknowledged, or recognised as needs, within a particular society. Marx's claim, then, is that under capitalism the socially acknowledged needs of workers extend only to the things that enable them to survive physically, in order that they may function as workers. In having their needs restricted in this way, Marx believes, workers are treated by capitalism as less than fully human. Marx finds this restricted view of workers' needs not only in the practice of capitalism, but recorded in the writings of political economists (Adam Smith in particular):

He [the political economist] tells us that originally and in theory the *whole product* of labour belongs to the worker. But at the same time he tells us that in actual fact what the worker gets is the smallest and utterly indispensable part of the product – as much, only, as is necessary for his existence, *not as a human being, but as a worker*, and for the propagation, not of humanity, but of the slave class of workers.[39]

Elsewhere in the *Manuscripts* Marx observes that, owing to the presence of an 'industrial reserve army' or 'relative surplus population' (as Marx was later to call the pool of unemployed workers), even the survival needs of workers receive only precarious recognition: 'Political economy . . . does not recognise the unemployed worker, the workingman, insofar as he happens to be outside [the] labour relationship. . . . For it, therefore, the worker's needs are but the one *need* – to maintain *him whilst he is working* and insofar as may be necessary to prevent the *race of labourers* from [dying] out.'[40]

In response to this, it may be noted that, especially in a modern technologically based society, the things that are needed in order to function

---

[38] A less concessionary but still plausible response to Benton's charge of 'speciesism' is offered by Lawrence Wilde (1998, pp. 134–5). He suggests that that Marx's account of animal needs is actually closer to the truth than Benton's, partly because Benton overestimates the degree of similarity between humans and animals implied by such things as rudimentary tool use by primates, and partly because Marx's account of animal needs is richer than Benton allows, including not only physical needs but also the need to hunt, to roam and to have companionship.

[39] *Economic and Philosophical Manuscripts*, p. 22; second emphasis mine. Though Marx attributes the views he outlines here generically to 'the political economist', he indicates that he has in mind particularly the writings of Adam Smith.

[40] *Ibid.*, p. 76. The terms 'industrial reserve army' and 'relative surplus population' are used in *Capital*, vol. I, ch. 23, sect. 3.

effectively as a worker may be rather more than the things needed for survival. Marx himself came to realise this, and in *Capital* he noted that the needs of the worker that capital must provide for in the form of wages include not only the means of survival, bodily replenishment and reproduction, but also a component relating to the education or training necessary to produce 'labour power of a developed and specific kind',[41] and further, that the 'so-called necessary requirements' of the worker contain a historical and moral element relating to 'the conditions in which, and consequently on the habits and expectations, with which the class of free workers has been formed'.[42] This modification to Marx's view amounts to a slight (though only a slight) softening of his indictment of capitalism: he is acknowledging that capitalism will sometimes – either because of its requirement for labour of a particular quality, or because of cultural conditions influencing the balance of power between workers and employers – pay wages to cover needs beyond mere survival and reproduction. This, however, does not affect the point being made here: that, for Marx, workers' socially acknowledged needs, insofar as they do approximate to the needs of survival, fall far short of what is needed for a fully human existence. This is borne out in the following quotation in which Marx, with his customary irony, sets out the means by which the capitalist – represented, as always, by the political economist – increases the satisfaction of his own needs at the expense of those of his workers:

(1) By reducing the worker's need to the barest and most miserable level of physical subsistence, and by reducing his activity to the most abstract mechanical movement; thus he says: Man has no need either of activity or of enjoyment. For he declares that this life, *too*, is *human* life and existence.

(2) By counting the most meagre form of life (existence) as the standard, indeed, as the general standard – general because it is applicable to the mass of men. He turns the worker into an insensible being lacking all needs, just as he changes his activity into a pure abstraction from all activity. To him, therefore, every luxury of the worker seems to be reprehensible, and everything that goes beyond the most abstract need – be it in the realm of passive enjoyment, or a manifestation of activity – seems to him a luxury. Political economy, this science of *wealth*, is therefore simultaneously the science of renunciation, of want, of *saving* – and it actually reaches the point where it *spares* man the *need* of either fresh *air* or physical *exercise*.[43]

---

[41] *Capital*, vol. 1, p. 276.   [42] *Ibid.*, p. 275.
[43] *Economic and Philosophical Manuscripts*, p. 104. A preliminary idea of what Marx thinks *is* needed for a human existence may be gained from the continuation of this passage, in which Marx continues to pillory the view of human needs presented by political economy: 'Self-renunciation, the renunciation of life and of all human needs, is its principal thesis. The less you eat, drink and buy books; the less you go to the theatre, the dance hall, the public house; the less you think, love, theorise, sing, paint, fence, etc., the more you *save* –

The message so far, then, is that Marx sets the level of human need higher than that which is necessary for bodily survival.[44] Human needs, for Marx, are not simply the conditions for the continued existence of human beings, but the conditions for their existence *as* human beings; the conditions for a recognisably human way of life.

We are now in a position to see how Marx's contrast between animal and human needs fits with the rest of his picture of human needs. Marx holds that the needs that, for workers, are suppressed under capitalism are those that are most distinctive of human beings. He therefore regards the worker's existence under capitalism as not only less than fully human, but also akin to that of an animal. This comparison has propaganda value for Marx, since in general people dislike the idea of being treated like animals. The comparison may also be informative to those who share, or at least understand, Marx's beliefs about the respective needs of animals and humans. To such people the comparison may serve as a shorthand way of communicating his beliefs about the difference between the range of needs that workers actually possess as human beings, and the range of needs attributed to them by capitalism. What must be emphasised, however, is that while Marx's comparison of human and animal needs may be helpful to him in explaining and disseminating his account of human needs and their distortion under capitalism, his theory does not depend on the accuracy of that comparison.

Consider, for example, the following passages:

It goes without saying that the *proletarian* ... is considered by political economy only as a *worker*. Political economy can therefore advance the proposition that the proletarian, *the same as any horse*, must get as much as will enable him to work. It does not consider him when he is not working as a human being; but leaves such considerations to criminal law, to doctors, to religion, to the statistical tables, to politics and to the poor-house overseer.[45]

[P]olitical economy knows the worker only as a working animal – as a beast reduced to the strictest bodily needs.[46]

The features of workers' socially acknowledged needs (i.e. those attributed to them under capitalism) identified here are those that we have already

---

the *greater* becomes your treasure which neither moths nor rust will devour – your *capital*. The less you *are*, the less you express your own life, the more you *have*, i.e., the greater is your *alienated* life, the greater is the store of your estranged being.'

[44] This point is also made in the *German Ideology*, where Marx writes that the proletarian 'is not in a position to satisfy even the needs that he has in common with all human beings ... his position does not even allow him to satisfy the needs arising directly from his human nature' (*MECW*, vol. v, p. 289).

[45] *Economic and Philosophical Manuscripts*, p. 24; second emphasis mine.   [46] *Ibid.*, p. 25.

encountered: that they amount only to what is needed as a condition for working or (what in Marx's day amounted to much the same thing) as a condition for survival, and that this is less than what is needed for a fully human existence. We will see shortly that the comparison with animal needs also pervades Marx's account of alienated labour, but there too everything that Marx says by means of the comparison he also explains in terms independent of his views on animal needs. Therefore, a reappraisal of animal needs (recognising perhaps that animals' needs, like those of workers, are less meagre than they are often taken to be) need not affect the substance of Marx's account either of the workers' socially acknowledged needs which he likens to the needs of animals, or of the human needs which he contrasts with both of these categories.

A preliminary conclusion that can be drawn from Marx's contrast between the needs that a worker has *qua* worker and *qua* human being is that he is committed to the expansion of needs, at least in the following sense. He is committed to the desirability of satisfying a larger set of needs than that which workers are acknowledged to possess under capitalism. This is confirmed in Marx's account of his socialist objective. He rejects a 'crude communism', motivated by envy and the 'urge to reduce things to a common level', as 'the abstract negation of the entire world of culture and civilisation, the regression to the *unnatural* simplicity of the *poor* and crude man who has few needs and who has not only failed to go beyond private property, but has not yet even reached it'.[47] This form of communism shares with capitalism its very narrow view of human needs, and therefore, Marx believes, offers no solution to the alienation of the worker from his human nature. Instead of this, Marx advocates a form of socialism characterised by the '*wealth* of human needs', a society populated by '*rich human being*[s]' possessed of '*rich human* need'.[48] In terms of the triadic analysis of need, A needs X for Y, Marx's claim is that capitalism and its ideology ascribe to workers only those needs for which Y ranges over survival and the ability to function as a wage worker, but that while such a conception of needs may be appropriate to other animals, there is for humans a wider range of ends which ought to be recognised as the basis for a correspondingly wider range of human needs.

We have seen that Marx is committed to some notion of the growth of needs. It appears, moreover, that the growth of needs required for Marx's vision of communism is not just a discrete step from one level of need-satisfaction to another, but an ongoing development. This can be discerned

[47] *Ibid.*, p. 89.   [48] *Ibid.*, pp. 101, 99.

from the fact that Marx criticises the crude, levelling-down version of communism not only for the low level of its aspirations but for their fixity – for setting itself 'a *definite, limited* standard'.[49] In order to judge the significance of this growth – whether it entails the permanent expansion of material production (as Heller, and many of Marx's green critics, suggest) and what are its likely ecological consequences – I will now examine further the nature of the expanded and expanding needs whose satisfaction Marx advocates.

6.4 Alienation and the needs of self-realisation

Two things are commonly said about the view of human needs in Marx's early works: first that he understands them as the exercise of essential human powers, and second that he understands them as social needs, for friendship, co-operation, etc. Both ideas can indeed be found in the *Manuscripts* and both are derived from what Marx takes (rightly or wrongly) to be distinctive features of human nature. Marx holds that it is part of human nature to engage in freely chosen, consciously pursued productive activity, and to relate, through that activity, to other humans and to the species as a whole.[50] In exercising their essential powers and entering into social relations humans can therefore be said to be realising their natures, and we may therefore refer to the corresponding needs as *needs of self-realisation*.[51]

One way of challenging this view of human nature would be to argue that the characteristics identified by Marx as distinctively human are to some extent shared by other animals. I have already shown, however, that this point can be conceded without otherwise altering Marx's account of human nature. A further and more powerful challenge would be to question Marx's reasons for putting productive activity at the centre of that account. Is this not, as his green critics are liable to argue, an unwarranted extrapolation from the ideology or practice of nineteenth-century industrialism, or an outmoded Enlightenment Prometheanism?

Marx does in fact give reasons for regarding productive activity as central, transhistorically, to human flourishing. He does so by identifying

---

[49] *Ibid.*, p. 89.   [50] *Ibid.*, p. 68.

[51] The exercise of essential powers and engagement in society may still be seen as realisation of human nature even if there are other species which to some degree share these characteristics. As Geras (1995, p. 48) notes, conceptions of human nature can include the characteristics shared by humans and by other species as well as the characteristics which distinguish humans from the others. For Marx's own use of the language of self-realisation, see, for example, text to note 64 below.

essential human powers, initially, with the human senses and other biolog-
ically founded (and hence transhistorical) human capacities: 'seeing,
hearing, smelling, tasting, feeling, thinking, observing, experiencing,
wanting, acting, loving'.[52] But these capacities, or 'human relations to the
world', do not remain static; they are not provided by nature in their final
form. Rather,

> the senses of the social man differ from those of the non-social man. Only through
> the objectively unfolded richness of man's essential being is the richness of subjec-
> tive human sensibility (a musical ear, an eye for beauty of form – in short, senses
> capable of human gratification, senses affirming themselves as essential powers of
> man) either cultivated or brought into being.[53]

Human self-realisation therefore requires not only the use of these natural
capacities, but their development. As the quotation indicates, the existence
of society is necessary to this development. It is necessary firstly insofar as
the capacities to be developed are directly social (love, for instance), and
secondly, even in the case of capacities that may be exercised in isolation,
because of the role that social phenomena like language or art play in their
development.[54] But what really drives the development of the human
powers, Marx thinks, what teaches us to use our senses to the full and provides us with new materials on which to exercise them, is the exploration
and manipulation of nature in productive activity. This is what he means
when he writes that 'the history of industry and the established objective
existence of industry are the open book of man's essential powers, the per-
ceptibly existing human psychology.'[55]

A proviso that Marx enters immediately is that though humans' essential powers have been developed through the history of their productive
activity, those powers have been exercised in an alienated way which has
actually inhibited the achievement of self-realisation. Science, transforming human life through the medium of industry, 'has prepared human
emancipation, though its immediate effect had to be the furthering of the
dehumanisation of man'.[56] What is needed then, for self-realisation, is free,
or unalienated productive activity. We might also wonder whether productive activity, or labour, is the only arena within which self-realisation can
be achieved. This is a question raised by Marx's tendency in his later works

---

[52] Economic and Philosophical Manuscripts, p. 94.
[53] Ibid., p. 96. Marx also writes that 'the human eye enjoys things in a way different from the
crude, non-human eye; the human ear different from the crude ear, etc.' (ibid., p. 95) and
that 'The forming of the five senses is a labour of the entire history of the world right down
to the present' (ibid., p. 96).
[54] Art is cited in the above quotation, language is mentioned in this role in ibid., p. 92.
[55] Ibid., p. 99.   [56] Ibid., p. 98.

to write of liberation being achieved through a reduction of necessary labour time, rather than a transformation of its character. In the *Manuscripts*, however, Marx assumes that self-realisation is to be achieved in the productive process, and for now I will leave this question open.

Marx's account of self-realisation is presented mainly by means of a contrast with alienated or estranged labour, in which human nature is *not* realised, but suppressed or distorted.[57] Alienated labour, according to Marx, is labour which 'is *external* to the worker', which 'does not belong to his intrinsic nature'. It is labour in which the worker 'does not affirm himself but denies himself, does not feel content but unhappy, does not develop freely his physical and mental energy but mortifies his body and ruins his mind'.[58] In part these descriptions refer to the material characteristics of alienated labour, resulting, for example, from the division of labour. The crucial point, however, concerns its intentional structure. Alienated labour, Marx continues, 'is . . . not voluntary, but coerced; it is forced labour. *It is therefore not the satisfaction of a need; it is merely a means to satisfy needs external to it.* Its alien character emerges clearly in the fact that as soon as no physical or other compulsion exists, labour is shunned like the plague.'[59] What Marx is arguing here is that labour can satisfy the worker's need to exercise his essential powers only if it performed at least in part with that intention, i.e. as an end in itself.[60] Wage labour, directed by another person and reluctantly performed, *solely* as a means for meeting basic, physical needs, fails to satisfy this criterion and therefore alienates the worker from his human nature. This does not mean that the satisfaction of basic needs can play *no* part in the motivation of unalienated productive activity; indeed, Marx's identification of *productive* activity as the 'life activity' in which human self-realisation is to be achieved appears specifically to allow (even to require) such an objective. What Marx appears to be saying then is that this objective cannot be the *sole* motivation; in order to achieve self-realisation by exercising his essential powers, the worker must want and intend *to produce* the objects to meet basic needs and not merely to *have* them.

[57] Chris Arthur (1986, p. 13) argues that in the *Manuscripts* the term 'labour' is reserved for productive activity under private property, whereas in Marx's later works it is used for productive activity as such. I will ignore this distinction and use the term in the more general sense.   [58] *Economic and Philosophical Manuscripts*, pp. 65–6.

[59] *Ibid.*, p. 66; my emphasis. See also p. 68: '*life activity, productive life* itself, appears to man in the first place merely as a *means* of satisfying a need – the need to maintain physical existence. Yet the productive life is the life of the species. It is life-engendering life. The whole character of a species – its species-character – is contained in the character of its life activity; and free, conscious activity is man's species-character. Life itself appears only as a *means to life.*'   [60] If this statement appears paradoxical, see note 68 below.

In 'On James Mill', a piece written around the same time as the *Economic and Philosophical Manuscripts*, Marx outlines the ways in which the wage-labour form alters both the material characteristics and the intentional structure of the labour process so as to reduce its capacity to bring about the worker's self-realisation:

> The product is produced for value, exchange-value, equivalency, and no longer because of its direct, personal connection with the producer. The more varied the production becomes, so the producer's needs are more varied while his activity becomes more one-sided and his labour can more and more be characterized as wage-labour, until finally it is purely this and it becomes quite accidental and inessential whether the producer has the immediate enjoyment of a product that he personally needs and also whether the very activity of his labour enables him to enjoy his personality, realize his natural capacities and spiritual aims.[61]

This passage is atypical, however, in its suggestion that self-realisation is to be achieved in producing for oneself. On the following page, Marx suggests that self-realisation is inhibited not only by the instrumental attitude to one's own activity that is fostered by wage labour but also by its self-interested motivation: 'Man – and this is the basic presupposition of private property – only produces in order to have. The aim of production is possession. Not only does production have this utilitarian aim; it also has a selfish aim; man produces only his own exclusive possession. The object of his production is the objectification of his immediate, selfish need.'[62] Marx is, of course, aware that in a market economy people produce not only for their own immediate need, but also for exchange; indeed, as the last quotation but one indicates, he believes that in a capitalist economy, production for the immediate use of the producer becomes increasingly marginal while production for exchange (commodity production) comes to dominate. Marx argues, however, that this form of production 'does not leave selfish need behind. It is rather an indirect way of satisfying a need that can only be objectified in the production of another and not in this production . . . I have produced for myself and not for you, as you have produced for yourself and not for me.'[63] The point here is that human nature is not only creative but social. Human self-realisation requires not only an ability to exercise one's powers creatively, but to do so for the benefit of fellow humans, thus expressing the species solidarity and need for recognition that Marx sees as part of human nature:

> Supposing that we had produced in a human manner; each of us would in his production have doubly affirmed himself and his fellow men. I would have: (1) objec-

---

[61] 'On James Mill', p. 118 (cf. alternative translation in *MECW*, vol. III, p. 220).
[62] *Ibid.*, p. 119.   [63] *Ibid.*

tified in my production my individuality and its peculiarity and thus both in my activity enjoyed an individual pleasure of realizing that my personality was objective, visible to the senses and thus a power raised beyond all doubt. (2) In your enjoyment or use of my product I would have had the direct enjoyment of realizing that I had both satisfied a human need by my work and also objectified the human essence and therefore fashioned for another human being the object that met his need. (3) I would have been for you the mediator between you and the species and thus been acknowledged and felt by you as a completion of your own essence and a necessary part of yourself and have thus realized that I am confirmed both in your thought and your love. (4) In my expression of my life I would have fashioned your expression of your life, and thus in my own activity have realized my own essence, my human, my communal essence.[64]

The view of human needs as including the need for appropriate forms of social interaction is found also in the *Manuscripts*, implicit in the discussion of alienation, where Marx explains how alienation, or estrangement, of the worker from his product and labour leads to the 'estrangement of man from man'.[65] The theme may also be detected in Marx's discussion of the contrasting significance of expanded needs under different social forms. Under socialism, as we have seen, Marx regards the satisfaction of expanded needs as 'a new manifestation of the forces of *human* nature and a new enrichment of *human* nature'; under a system of private property, however, 'every person speculates on creating a *new* need in another' in order 'to place him in a new dependence' and 'thereby to find satisfaction of his own selfish need'.[66] The main significance of this passage – discussed above – is its reference to false or 'manipulated' needs, but it also reinforces the thought that, just as the human need to express one's creative powers is alienated and perverted when it becomes a mere means to the satisfaction of one's physical needs, so the need for social intercourse is perverted when it becomes merely a means to the satisfaction of individual needs. Marx believes that the need for social intercourse as an end in itself can develop fully only with the abolition of private property, but is prefigured in the relations among communist workers:

When communist *artisans* associate with one another, theory, propaganda, etc., is their first end. But at the same time, as a result of this association, they acquire a new need – the need for society – and what appears as a means becomes an end. In this practical process the most splendid results are to be observed whenever French socialist workers are seen together. Such things as smoking, drinking, eating, etc., are no longer means of contact or means that bring them together. Association, society and conversation, which again has association as its end, are enough for them; the brotherhood of man is no mere phrase with them, but a fact

---

[64] *Ibid.*, p. 122.   [65] *Economic and Philosophical Manuscripts*, p. 69.   [66] *Ibid.*, p. 101.

of life, and the nobility of man shines upon us from their work-hardened bodies.[67]

The purpose of this extended exploration of Marx's concept of alienated labour has been to show that the needs of self-realisation are at the heart of Marx's understanding of human need as it is developed in the *Manuscripts*.[68] Marx's commitment to a growth of needs beyond those that are necessary for physical survival should be interpreted in the light of this fact; and in this light his insistence that a genuinely human existence requires the satisfaction of needs over and above the 'basic' needs recognised by political economy appears to be a matter not so much of *what* should be produced as *how* productive activity should be conducted. Marx's resistance to an excessive emphasis on the satisfaction of basic needs, therefore, arises not so much from a wish for increased material consumption as from concern about the distortion of human nature that occurs when human activity becomes a mere means to the satisfaction of these 'animal' needs.

That is not to say, however, that the Marx of the *Manuscripts* is unconcerned with physical need. In chapter 4 I argued that, at least from the *Manuscripts* onwards, Marx endorsed what I called the Principle of Ecological Dependence, premised upon his materialist view of humans as beings with physical needs that must be met if they are to survive and as a condition for anything else that they may do. That the 'realisation of labour', the exercise of human powers, is a physical process in which labour 'realises', or 'objectifies', itself by creating objects that are necessary both for the worker's survival and as conditions for further creative activity is clear in the following comment, in which Marx notes the possible

---

[67] *Ibid.*, pp. 109–10.

[68] We can express this in terms of the triadic analysis of need. That analysis states that all needs have the form '*A* needs *X* for *Y*', and Marx's point is that in the ideology and practice of capitalism *Y* ranges over physical survival or the ability to function as a worker, an interpretation which he regards as too narrow since important normative force attaches also to the needs specified by substituting 'self-realisation' or 'flourishing' for *Y*. This formulation on its own is not very informative, but Marx fleshes it out by identifying certain forms of creative activity and social interaction as the things (*X*s) we need to do in order to flourish or 'realise ourselves'. An interesting point is raised here by a disanalogy between survival needs and needs of self-realisation. In the case of survival needs the means–end relation between *X* and *Y* is purely causal, so it is not necessary to identify the appropriate means in order to grasp what a survival need is. In the case of needs of self-realisation, however, the means – exercise of powers, etc. – appear to be partly constitutive of flourishing, or self-realisation, hence the sense that the concept 'needs of self-realisation' is not adequately specified until the means have been identified. The fact that the means are partly constitutive of the end in this case explains how it is possible both to exercise one's essential powers as an end in itself and to do so as a means to self-realisation. Cf. text to note 60 above.

consequences for the worker of being alienated from his product: 'So much does labour's realisation appear as loss of realisation that the worker loses realisation to the point of starving to death. So much does labour's objectification appear as loss of the object that the worker is robbed of the objects most necessary not only for his life but for his work.'[69]

And it is not only as conditions of life that material things are necessary to the exercise of human powers. We saw in section 4.6 that Marx's analysis of the labour process recognises the roles played by material objects, as *objects of labour* (raw materials) and as *instruments of labour* (tools). The need for material things as objects upon which to exercise human powers is acknowledged in the following passage:

> *Man* is directly a *natural being*. As a natural being and as a living being he is on the one hand endowed with *natural powers, vital powers* – he is an *active* natural being. These forces exist in him as tendencies and abilities – as *instincts*. On the other hand, as a natural, corporeal, sensuous, objective being he is a *suffering*, conditioned, and limited creature, like animals and plants. That is to say, the *objects* of his instincts exist outside him, as *objects* independent of him; yet these objects are *objects that he needs – essential objects, indispensable to the manifestation and confirmation of his essential powers.*[70]

So humans need material resources in order to sustain (i) their lives and (ii) their self-realising activities. These two reasons are explicitly noted and distinguished by Marx in the following passages from the *Manuscripts*. Firstly, he notes that 'nature provides labour with [the] *means of life* in the sense that labour cannot *live* without objects on which to operate', as well as providing 'the *means of life* in the more restricted sense, i.e., the means for the physical subsistence of the *worker* himself'; and secondly, in connection with his famous reference to nature as man's 'inorganic body', Marx writes that 'nature is (1) his direct means of life, and (2) the material, the object and the instrument of his life activity'.[71]

The fact that material objects are needed as means for the exercise of human powers is of some importance as it suggests that the *development* of those powers, advocated by Marx as a component of human self-realisation, may result in an *expansion* of material needs. Consider some of the ways in which people seek to exercise their human powers. Even those that appear modest in their requirements may become less so as the agent's skill and ambition increase and as technology brings new possibilities into view. The gardener may begin with a small plot of land, but if his ambitions grow with his skill (and others' do likewise) then the countryside will

---

[69] *Economic and Philosophical Manuscripts*, p. 63.
[70] *Ibid.*, p. 136; final emphasis mine.   [71] *Ibid.*, pp. 64, 67.

soon disappear under suburban sprawl. The violinist, having progressed to a Stradivarius, may feel that the only way in which his skills may be further advanced is by acquiring a recording studio or concert hall; and the cyclist may find that her performance just cannot be further enhanced on a bike built from anything less than titanium (ultra-light but energy-intensive to produce). And similarly the photographer's self-realisation would be limited with a Box Brownie, the tennis player's with a wooden racket, and so on. And what if I feel that my capacities are best exercised by competing in car races or flying aircraft . . .? And if we assume (as Marx does in the *Manuscripts*) that self-realisation is to be achieved at least partly within the production process, then Marx's description of the history of industry as 'the *open* book of *man's essential powers*'[72] may suggest that workers will achieve maximal levels of human fulfilment, exercising their creative powers and at the same time expressing their human solidarity, by applying their ingenuity and skill to the expansion of industrial output.

What has been suggested here is that, given (i) Marx's commitment to the development of human needs of self-realisation, and (ii) the necessity of material objects as conditions for the activities that constitute human self-realisation, there is reason to think that Marx may also be committed to (iii) the growth of these material needs. This, however, need not follow. For although the exercise of essential human powers involves the use and consumption of material objects, the *development* of those powers need not entail the consumption of *more* objects as a means of achieving more of whatever is the goal of the activity. Within the production process, *development of human powers* could as well be manifested in the use of human ingenuity to achieve as much or more with fewer resources or to increase what can be achieved at existing levels of resource consumption. This point recalls the one made earlier, that the development of technology can be used not only to increase the volume of production but also to increase the efficiency of resource use, so that (within limits) existing or increased levels of production can be achieved with lower environmental impact. It is surely true that the application of human ingenuity to this objective – provided it is carried out under appropriate non-alienating conditions, such that the agent's ends are the ends internal to the activity, etc. – qualifies as a development of essential human powers.[73] And similarly in the case of

---

[72] *Ibid.*, p. 97.
[73] This claim fits well with the view of human nature that (as outlined in the last chapter) Cohen ascribes to Marx, according to which it is in the nature of humans to be ingenious enough to devise solutions to their problems, and rational enough to retain those which work. A possible objection (suggested by Andrew Chitty) is that the application of human ingenuity to controlling or reducing the environmental impact of human activity fails to

the 'non-productive' activities considered in the last paragraph, human powers may be exercised and indeed developed in striving to achieve as much as possible within certain constraints. This is what happens, for example, when the rules of a sport restrict the specifications of equipment to be used – such rules are not usually thought of as preventing the development of human powers but as channelling that development in certain directions rather than others. Provided that the rules have a rationale which can be accepted by participants there is no reason to think that the existence of the rules undermines their ability to realise themselves in that activity.

What the above argument shows is that the needs of self-realisation (the expanded component of human needs endorsed by Marx in the *Manuscripts*) can be satisfied in ecologically benign as well as ecologically damaging ways. It appears then that the structure of such needs is disjunctive – people need *either* to exercise their powers in benign ways, *or* to exercise their powers in damaging ways, in order to achieve self-realisation – and that a society which provides the means for the exercise of human powers in one or the other of these ways will be a society in which human needs can be satisfied.[74] If this is correct, then Marx's commitment to the satisfaction of human needs would appear not to commit him to a (potentially) ecologically damaging expansion of material production. This, it seems to me, is the correct conclusion to draw as far as Marx's conception of needs in the *Economic and Philosophical Manuscripts* is concerned. Before reaching this conclusion, however, there is an objection which needs to be considered.

---

satisfy what may be termed the 'productivism' inherent in Marx's account of self-realisation – that is, his view that self-realisation involves the *objectification* of an agent's personality through the use of his powers to create new objects that will satisfy the needs of fellow humans (see, for example, 'On James Mill', p. 122, quoted above). However, I am not suggesting that such an application of human ingenuity is an *alternative* to the production of material objects, which of course are essential to human survival. The suggestion, rather, is that the satisfaction derived from such production need not depend only upon the quantity of the objects produced, nor even upon their intrinsic qualities, but may result also from extrinsic or relational properties such as the relative absence of environmentally damaging side-effects in their production or use. Indeed, it is plausible to suppose that Marxian self-realisation *requires* the minimisation of such effects, since they would otherwise tend to undermine the satisfaction of human needs which for Marx is the purpose of self-realising activity.

[74] Alternatively, as the previous footnote suggested, it may be that self-realisation can *only* be achieved by the exercise of human powers in ecologically benign ways. In what follows, however, I will assume only the weaker, disjunctive claim.

6.5 The free-choice objection

Against the conclusion just outlined, it might be objected that a society which provided for human powers to be exercised only in ecologically benign ways would fail to satisfy the requirement that self-realising activity be freely chosen, and would therefore not be a society in which fully human needs could be satisfied.[75] If this is correct then Marx is faced with a dilemma: either he advocates unrestricted choice in the exercise of human powers, in which case some people will choose to exercise their powers in ways which, through the production of ecological problems, affect the ability of others to satisfy their needs; or he restricts the ways in which human powers may be exercised, denying free choice and thus undermining the ability of those who are subject to the restriction to satisfy their needs of self-realisation. I suggest, however, that this objection depends upon a misinterpretation of the freely chosen character of self-realising activity. The objection assumes that, in order to satisfy her needs of self-realisation, an agent must have absolute freedom to choose her forms of activity, including the freedom to choose forms which interfere with the need-satisfaction of others. Such an interpretation is not warranted by what Marx says about the free character of self-realising activity, and cannot reasonably be imputed to him in the absence of such textual evidence.

Free human activity, for Marx, is to be contrasted, firstly, with the activity of animals, and secondly, with alienated activity which he regards as akin to that of animals. The salient point here is that on Marx's view an animal's activity involves *no* element of choice; it is completely determined by:

(i)   its immediate physical needs: it 'produces only what it immediately needs for itself or its young'; and

(ii)   the instincts of its species: it 'forms objects only in accordance with the standard and the need of the species to which it belongs'.

Humans on the other hand do have choices:

(i')   they are able to produce more than enough to satisfy their most basic physical needs and can therefore choose how to spend their remain-

---

[75] The premise of this objection – that self-realising activity must be, in a strong sense, freely chosen – appears to be endorsed by Elster (1985, p. 524). Elster is not explicitly addressing ecological issues, but argues that such freedom would result in some individuals choosing expensive activities in which to realise themselves, resulting in a lowering of the general level of self-realisation. See also Archard 1987, pp. 21, 25.

ing productive time: 'man produces even when he is free from physical need and only truly produces in freedom therefrom'; and

(ii') the intelligence which enables them to produce more than they immediately need also frees them from instinct and enables them to find alternative ways of satisfying their needs: 'man knows how to produce in accordance with the standards of every species, and knows how to apply everywhere the inherent standard to the object. Man therefore also forms objects in accordance with the laws of beauty.'[76]

So the thing that distinguishes truly human production from animal-like alienated production, in Marx's view, is not an absolute freedom to do whatever one likes, but an ability to exercise some control over the aims and methods of one's creative activity. This control, Marx believes, is lost in alienated labour in which the aim and the method of production are determined not by the worker but by his employer. Marx's account of alienated labour also suggests another element of the freedom of self-realising human activity. Alienated labour, he notes, is coerced in that it is conducted under the pressure of physical necessity, as a means of obtaining a wage. What he is suggesting, however, is not that people should be able to choose not to work – such a choice could only be a minority privilege since physical needs have to be met – but that labour should become an end in itself, and so no longer coerced. 'Free' productive activity, then, is for Marx activity in which producers can exercise a degree of control over what and how they produce, and in which they can be motivated by the activity itself and its internal goals, and not solely by external considerations. Such freedom does not require the ability to engage in whatever forms of activity one likes, and would not be undermined by the fact that a society provided for the exercise of human powers only in ecologically benign ways.

This interpretation is supported by the following more general considerations. Norman Geras, defending Marx against charges of utopianism in his notion of abundance, argues that it would be methodologically wrong, in the absence of decisive textual evidence, to impute to Marx an obviously absurd belief. The suggestion that under socialism everybody could have whatever they might feel themselves to need as means of self-realisation would be such an absurdity, since it is obvious (even leaving aside ecological factors of which Marx might not have been aware) that different people's perceived needs may come into conflict. As Geras argues, violins and bicycles for those who feel themselves to need them may be all right,

---

[76] *Economic and Philosophical Manuscripts*, pp. 68–9.

but a perceived need for huge tracts of land (the size of Australia) in which to wander undisturbed, even if shared by only a small part of the population, obviously will not.[77] It is therefore inevitable that there will be limits to what one may do in pursuit of one's free self-realisation, and in the absence of evidence to the contrary we should assume that Marx recognises this fact.

It may be objected that the sort of limit implied by Geras's example is far above the sort of limit that must be regarded as necessary given what we now know about ecological scarcity, and that the sorts of resources for self-realising activity now deemed ecologically unsustainable (such as the private planes, recording studios etc. referred to in the last section) would not have appeared unreasonable to Marx, given the technological optimism of his time. But, even if this objection is conceded, there is another much more immediate set of limits on an individual's self-realising activity which Marx must have recognised. Self-realisation, for Marx, is to be achieved at least in part through participation in co-operative, social activity, and will therefore be subject to the constraints on the individual that are involved in any form of co-operation. This would have been clear to Marx, and we should therefore avoid imputing to Marx a concept of self-realising activity as free of all constraint, unless textual evidence forces us to do otherwise. Since we have already seen that Marx's description of self-realising activity does not require it to be free of all constraint, we must reject the premise upon which the free-choice objection is founded.

What the above argument shows is that the mere existence of constraints upon the ways in which human powers may be exercised is not sufficient to undermine the possibility for self-realisation. We may also note that, in their general structure, constraints on self-realising activity designed to control its ecological consequences are no different from constraints required by social co-operation in production. Both are constraints upon the ways in which individuals may go about satisfying their needs of self-realisation, motivated by more general considerations of human need-satisfaction (in one case the needs that will be satisfied by the products of co-operative labour, and in the other the needs that will go unsatisfied as a result of ecological problems). So, if co-operative constraints are compatible with free self-realising activity, as we must suppose Marx to have believed, then there is every reason to suppose ecological constraints also to be compatible – as indeed the examination of Marx's account of 'free' self-realising activity suggested.

[77] Geras 1989, p. 80.

6.6 Needs in Marx's later works

I have tried to show that Marx's *Economic and Philosophical Manuscripts* provide a conception of human need in which the satisfaction of human needs is not bound to have deleterious ecological consequences. It is sufficient for my purposes that there is a plausible conception of human need which is ecologically sustainable in this way and which can be incorporated into the interpretation of historical materialism offered in the last chapter. While I have indicated (in the text and footnotes) that there are passages from Marx's later works which tend to support the account of human need developed in the *Manuscripts*, I am doubtful whether the comments on needs contained in these later works are sufficient to determine conclusively whether Marx continued to adhere to the *Manuscripts'* conception of human need. My claim, therefore, is that Marx *could* have done so, and that this makes possible an ecologically sustainable historical materialism.

Even though I am not claiming to prove a continuity in Marx's understanding of human need, some examination of his later comments on needs remains necessary, firstly because they may reveal ways in which his historical materialism commits him to a view of needs different from that described above, and secondly because they may help to clarify and refine that account. I will therefore look briefly at some of the ways in which Marx's later works might be thought to exhibit a different conception of need from that contained in the *Manuscripts*.

One aspect of Marx's later works which may appear to indicate such a difference is his enthusiasm for the growth of needs that capitalism promotes. It will be recalled that in his earlier works Marx recognises that capitalism has such an influence, but emphasises that the 'needs' arising from it may in fact be false needs, incompatible with the rich and many-sided needs required for human flourishing. In the *Grundrisse*, however, Marx links this growth of (perceived) needs with the historically progressive character of capitalism, which he applauds in a manner similar to that for which the *Communist Manifesto* is famous:

capital has pushed beyond national boundaries and prejudices, beyond the deification of nature and the *inherited, self-sufficient satisfaction of existing needs* confined within well-defined bounds, and the reproduction of the traditional way of life. It is destructive of all this, and permanently revolutionary, tearing down all obstacles that impede the development of productive forces, *the expansion of needs*, the diversity of production and the exploitation and exchange of natural and intellectual forces.[78]

[78] This passage quoted from the translation in *KMSW*, p. 364; my emphasis.

Statements like this may appear to indicate an abandonment by Marx of his earlier distinction between true and false needs, and an embracing of the development of new appetites irrespective of their content. Such an interpretation, however, cannot be sustained. I have already noted how in *Capital* Marx contrasts use-value production, aimed at the satisfaction of human needs, with commodity production, aimed at the production of exchange value and indifferent to whether the appetites that it serves are genuine human needs. Marx makes a similar move in the *Grundrisse*, contrasting 'the old view [of antiquity], in which the human being appears as the aim of production' with 'the modern world, where production appears as the aim of mankind and wealth as the aim of production'.[79] He argues that while capitalism does produce a development of 'needs, capacities, pleasures, productive forces etc.', it does so in a distorted and alienated form: 'In bourgeois economics – and in the epoch of production to which it corresponds – this complete working-out of the human content appears as a complete emptying-out, this universal objectification as total alienation, and the tearing down of all limited, one-sided aims as sacrifice of the human end-in-itself to an entirely external end.'

What is clear from these passages is that while in general terms Marx welcomes the expanding perception of need which capitalism brings about, he remains critical of the particular forms that it takes. Further investigation reveals that Marx's reasons for welcoming the process despite its unsatisfactory form have to do with the same features of human nature that he thought important in the *Manuscripts*. Marx values the expansion of human appetites primarily for the development of human powers that this implies, and he values the development of human powers not simply as a means to increased consumption but as an end in itself, a component of human flourishing: 'Capital's ceaseless striving towards the general form of wealth drives labour beyond the limits of its natural paltriness, and thus creates the material elements for the development of the rich individuality whose labour also therefore appears no longer as labour, but as the full development of activity itself . . .'.[80]

This account corresponds closely to Marx's earlier account of human needs as the exercise of essential human powers, and in particular to his account of the history of industry as the open book of man's essential powers.[81] However, Marx does not regard the increase of human powers only as an intrinsic value; he also values it instrumentally, for the opportunity it creates for a reduction in labour time and the elimination of monot-

---

[79] *Grundrisse*, pp. 487–8.   [80] *Ibid.*, p. 325   [81] *Economic and Philosophical Manuscripts*, p. 99.

onous and unnecessary work. Capitalism, he writes, will have achieved its historic destiny when

the development of the productive powers of labour, which capital incessantly whips onward with its unlimited mania for wealth . . . [has] flourished to the stage where the possession and preservation of general wealth require a lesser labour time of society as a whole, and where the labouring society relates scientifically to the process of its progressive reproduction, its reproduction in a constantly greater abundance; hence where labour in which a human being does what a thing could do has ceased.[82]

This passage relates to an aspect of the development of Marx's thought alluded to earlier. In his later works, Marx appears in places to have abandoned his earlier view that self-realisation is to be achieved primarily within the production process, and instead looks forward to the lessening of necessary labour time as a means of enabling self-realisation to be achieved *outside* the labour process. It has to be said that Marx's reasoning about this is unclear. In *Capital*, for example, he indicates that he still wishes to see necessary labour conducted in ways more appropriate to human nature, but now anticipates limits to the degree to which this can be achieved:

Freedom, in this sphere, can consist only in this, that socialized man, the associated producers, govern the human metabolism with nature in a rational way, bringing it under their collective control instead of being dominated by it as a blind power; accomplishing it with the least expenditure of energy and in conditions most worthy and appropriate for their human nature. But this always remains a realm of necessity. The true realm of freedom, the development of human powers as an end in itself, begins beyond it, though it can only flourish with this realm of necessity as its basis. The reduction of the working day is the basic prerequisite.[83]

Marx's reason here, for seeing only a limited potential for the achievement of self-realisation within the production process, is that activity within the production process cannot be absolutely free, being constrained by the needs which it serves. However, I have already shown (in the discussion of the 'free-choice objection') that absolute freedom of the kind referred to here is not necessary for the exercise of what Marx identifies as essential human powers, and that Marx implicitly and sometimes explicitly acknowledges that self-realising activity will involve constraints upon an individual's freedom. Marx has therefore not shown by this argument that self-realisation is unachievable within the production process.

There may of course be other reasons for thinking that the exercise of human powers is achievable only to a limited degree within the

---

[82] *Grundrisse*, p. 325. [83] *Capital*, vol. III, p. 959.

production process,[84] but this is really beside the point. The point is that, as I have argued, Marx's commitment to growth of needs refers primarily to the needs of self-realisation, and the satisfaction of such needs, whether within or outside the production process, does not necessitate the expansion of resource consumption and the concomitant exacerbation of ecological problems. Rather, the achievement of self-realisation will depend upon the minimisation of such problems, and may even be effected through activities aimed at bringing about that minimisation. That is not to deny that Marx would advocate using resources, if available, to increase material consumption. His preferred, ideal scenario would be one in which increased material consumption *and* reduced working time could be achieved, while at the same time making the remaining necessary labour time more fulfilling. Since Marx believed something like this to be possible it would be inaccurate to portray him as a green theorist ahead of his time. What needs to be considered, however, is how Marx could or would have responded to a situation in which scarcity of resources (taken here in a wide sense which includes the capacity of the ecosystem to absorb pollutants) renders his ideal scenario impossible and requires a choice to be made between its different elements. My contention in this chapter is, firstly, that it is implicit in Marx's account of human needs that he would prioritise those which relate most closely to human flourishing, and secondly that, given his account of human flourishing and the emphasis it places upon creative and social self-realising activity, he could make such a prioritisation, dropping the aspiration for greater consumption to the extent that ecological scarcity requires, without abandoning the goal of a reasonable level of self-realisation for all human beings.

6.7 Conclusion

In this chapter I have attempted to show that Marx's commitment to the satisfaction of human needs, and moreover to the satisfaction of expanded human needs, can be interpreted in a way which does not entail an increase in the ecological impact of human activity. One way of doing this would be to interpret Marx's use of the communist slogan 'To each according to his needs!' not as a commitment to the satisfaction of all needs or to any particular level of need-satisfaction, but as a criterion for the just distribution of goods – those with the greatest need should receive the most. This strategy, however, is untenable, for when Marx uses the concept of need in

---

[84] See, for example, the discussion of technological alienation in Grundmann 1991b.

the *Economic and Philosophical Manuscripts* as part of his critique of capitalism it is clear that he objects not merely to the maldistribution of goods but
to the absolute insufficiency of need-satisfaction that characterises the lives
of workers under capitalism; his goal is not just the more equal sharing of
benefits and burdens but achievement of fulfilled and non-alienated lives.
This explains Marx's rejection in the *German Ideology* of the idea of imple-
menting the communist slogan without also achieving a minimum level of
need-satisfaction: far from ensuring human fulfilment, this would merely
cause *want* to be made general.[85] Marx also has an instrumental reason for
requiring some minimum level of need-satisfaction before the communist
slogan can be implemented. Not only would the generalisation of poverty
fail to satisfy the value that Marx's use of the slogan is intended to express;
it would make the slogan unsustainable even as a pattern of relative distribution, since the pattern would collapse in the ensuing 'struggle for
necessities'.[86]

So Marx is committed, both as a part of his goal and as a necessary means
to the achievement of that goal, to the achievement of some minimum level
of need-satisfaction, and it is this which in turn necessitates the 'greater
abundance' to which he refers. But he is not committed to the satisfaction
of *all* needs. Indeed, it is hard to see what such a commitment could
amount to, since even if we consider only 'true' needs it is easy to see that
there may be many different goods each of which would genuinely contribute to a person's well-being but which are not jointly attainable. The
task of this chapter has therefore been to consider what sort of level of
need-satisfaction is implied in Marx's use of the communist slogan and his
other references to human needs. Talk of 'levels', however, may be misleading if it is taken to imply that this is a purely quantitative rather than
a qualitative matter. I have argued that Marx understands needs in terms
of the conditions for human flourishing, but since there are many ways in
which Marx's conception of a flourishing human life might be realised, it
is not possible to give a simple list of needs understood as the necessary
means to that end. Instead, I have focused upon the end – Marx's conception of human flourishing, or self-realisation – and argued that among the
possible ways in which it may be realised are ways which do not depend
upon the consumption of vast or ever-increasing quantities of natural
resources.

I do not say that a Marxian conception of human need-satisfaction can
be achieved without *any* increase in material consumption, for although

---

[85] *The German Ideology*, p. 56 (see text to note 64, ch. 5 above).   [86] *Ibid.*

his account of self-realisation is flexible as regards its material form and therefore its resource requirements, self-realisation does presuppose a reasonable satisfaction of more 'basic' needs – for food, shelter, education and so on – and given the continued deficiency of these goods in many parts of the world it cannot be maintained that Marxian abundance has been achieved. Marx's ideal of universal human flourishing would require an expansion in the supply of at least *these* goods, in the relevant parts of the world. In order to avoid thereby exacerbating ecological problems and undermining these or other needs, the increased supply of basic goods would have to be achieved by some combination of redistribution and (assuming this to be insufficient) technological innovation.[87] Marxists therefore cannot dismiss technological innovation as a means of avoiding or mitigating ecological problems. Many greens will condemn this as an example of the 'technological fix' approach to ecological problems. This, however, is too strong. Unlike some other targets of green criticism, and contrary to what some greens maintain, Marxists are not committed to an unlimited technological amelioration of ecological problems. Certainly their conception of human flourishing will lead them to seek technological means of increasing the 'ecological efficiency' of production wherever possible, in order to maximise the possibilities for human flourishing while minimising its ecological impact. All that they strictly *require* from technological development, however, is the finite and therefore much more feasible amelioration that will be necessary in order to make the basic preconditions for human self-realisation available to all.

[87] On the question of whether increased production is necessary in order to meet basic needs on a global scale, see the work of Susan George and Amartya Sen, referred to in section 2.4.

Conclusion

What I have argued in this book is that, contrary to the contentions of many environmental commentators, the main commitments of Marx's theory of historical materialism can plausibly be interpreted in a manner that is compatible with a recognition of environmental problems and constraints. I have suggested, moreover, that Marx's account of the ways in which human societies depend upon and are influenced by their natural environments, both at a macro-level (historical materialism) and at a micro-level (conception of the labour process), offers a helpful framework within which to investigate the causes and solutions of ecological problems.

At the heart of my argument is what may be termed a qualitative interpretation of Marx's notion of the development of the productive forces. According to this interpretation, development of the productive forces is not a unilinear or one-dimensional process. There is no single path that it must follow, and no single substantive criterion against which all such developments can be measured. The one feature that all developments of the productive forces share, and which qualifies them as such, is that in some way they contribute to the solution of problems faced by human beings or to the furtherance of their interests; but these interests and problems vary according to material and social circumstances and between different groups in society. My critique of Cohen, it will be recalled, was that while he was right to see the problem-solving capacity of human beings as (ultimately) the motive force behind the development of the productive forces, he was mistaken to assume that only one problem – the need to increase labour productivity – motivates the exercise of that capacity. Since the problems faced by human beings include – increasingly – ecological problems, the solution or amelioration of such problems can be included among the criteria for development of the productive forces. This account, I have argued, is consistent with the role played by the development of the productive forces in Marx's theory, including that of making possible the satisfaction of needs that he sees as a precondition for the establishment of communism.

The idea that the development of the productive forces is to be interpreted qualitatively is also to be found in Reiner Grundmann's interpretation of Marx, an interpretation which is, however, in other respects rather different from mine. I hope, therefore, that a brief examination of some of these differences will help to clarify the substance and implications of my own account.

Grundmann in fact interprets the term 'growth of the productive forces' as having a 'double meaning' for Marx:

it can mean (1) increasing mastery over nature and (2) production of wealth (material goods) with ever-diminishing effort or in increasing abundance. The first meaning is that mankind gains an ever-increasing mastery over nature, in the sense that individuals develop into universal human beings, that they expand their control over the world around them, that they are able to shape a world according to their needs and pleasures. . . . The second meaning is primarily economic; a growth in this sense can be measured with the economic criteria of efficiency.[1]

The first meaning corresponds roughly to the qualitative account of the development of the productive forces that I have offered, especially in its emphasis on the need of humans to exercise control over their surroundings, while the second meaning coincides with Cohen's quantitative interpretation of that development as a development of labour productivity. Grundmann is right to remark that for Marx these two meanings are linked, in that '[t]he dignity of human beings requires freedom from hunger as much as it does freedom from a hostile nature which acts upon them as an alien force'.[2] I would suggest, however, that the reference to a double meaning is liable to be misleading. As I expressed it in chapter 6, the satisfaction of basic physical needs is a necessary though not sufficient condition for human self-realisation, and the development of the productive forces must therefore include such developments of labour productivity as are necessary to permit the satisfaction of these basic needs. This, however, does not constitute a dual meaning; rather, productive development in the narrow sense should be seen as one element (or component, or dimension) of productive development in the broader sense. Marx's mistake was not to equivocate between two distinct meanings but to write at times as if the narrower, economic form of productive development exhausted the range of that concept. On the whole, however, Grundmann

---

[1] Grundmann 1991b, p. 4.   [2] *Ibid.*

and I are in agreement in placing the broad, qualitative interpretation of productive development at the heart of Marx's account.

Grundmann contrasts this broad, qualitative account of productive development with the accounts of some of Marx's followers. He writes:

> I interpret Marx's statement that the productive forces develop throughout history and that they must be unfettered if social relations impinge upon them, in the wide sense which I explained a moment ago as a process of unfolding human self-realization. Orthodox Marxism has always interpreted this statement from the 1859 *Preface* in a narrow economic sense.[3]

This, however, is where Grundmann's account and mine begin to diverge. While I am happy to accept this characterisation of orthodox Marxism (at least for the purposes of this argument), it seems to me that Grundmann's rejection of Marxist orthodoxy, and indeed of central tenets of Marx's own position, is far more sweeping than the adoption of a qualitative interpretation of the development of the productive forces requires. The central issue on which Grundmann departs from orthodox Marxism, and from Marx's own view, is the significance of the relations of production. Grundmann acknowledges that Marx was inclined (for example in his comments on soil fertility) to view capitalist relations of production as the main cause of ecological problems, but he regards this as a mistake,[4] claiming that, on the contrary, 'a socialist society is in no *structurally* better position [than capitalism] to avoid ecological problems'.[5]

What leads Grundmann to make this claim? Several of his reasons relate to the experience and particular features of historical Soviet-type societies, but as Grundmann recognises, other models of socialist society may be constructed which do not share these features.[6] Grundmann also argues that collective ownership makes socialist societies vulnerable to the 'tragedy of the commons' and other problems of public goods and externalised costs, but this argument too depends upon the particular structures advocated, and these would have to be elaborated and analysed in order to make Grundmann's argument effective. Grundmann's claim that the phenomenon of unforeseen and therefore unintended consequences undermines planning as a solution to ecological problems is vulnerable to the same objection, insofar as the structuring of agents' interests affects the range of consequences that are foreseen. To the extent that there are

---

[3] *Ibid.*, p. 54. See also pp. 235–6 on fettering.
[4] *Ibid.*, p. 74: 'I plead for a reinterpretation of Marx in this respect which acknowledges frankly Marx's own predominant approach (i.e. blaming capitalism's social form) but does not accept it as the main tool in analysing contemporary ecological problems.'
[5] *Ibid.*, p. 43.   [6] *Ibid.*, pp. 41–3.

consequences which cannot be foreseen at all, this factor is neutral between different structures but does not preclude social structure making a difference to those ecological problems that are foreseeable. Taken together these arguments discredit the simplistic views that socialist relations of production are in themselves sufficient to ensure a better handling of ecological problems, and that changes in social structure can eliminate ecological problems altogether, but they do not refute the more modest claim that changes in social structure can make a difference.

These, however, are not the main reasons why Grundmann rejects Marx's emphasis on social relations. Grundmann's main reason derives from his claim (discussed in section 5.2 above) that modern technology, in virtue of its complexity and tight-coupling, has an inherent tendency to produce ecological problems, independently of the social structures within which it is used. Parallel with this is Grundmann's claim that complex modern technology is inherently alienating, because it deprives workers of their skills while obstructing their understanding of the productive process.[7] Grundmann thus concludes that the solution to both ecological problems and alienation depends not upon social structure but upon the possibility of exercising control over technological development.[8]

I suggested in chapter 5 that Grundmann may have overstated the inevitability of modern technology producing ecological problems, and similar doubts may be raised regarding its connection with alienation.[9] In any case, Grundmann's argument does not support his contention that socialism is in *no* better a position than capitalism to deal with ecological problems, since not all technologies have the characteristics that supposedly produce the problems. But suppose, for the sake of the argument, we concede Grundmann's claim that changes in social structure are unable to affect the ecological consequences of existing technologies. This still leaves open the possibility, suggested by my interpretation of historical materialism in chapter 5, that social structure may affect the trajectory of technological development, so as to make it more or less likely that technologies capable of improving efficiency of resource use, reducing pollution, etc. will be developed and put into operation. If so, then social structure remains – for reasons both of ecology and alienation – a much more significant political issue than Grundmann acknowledges.

---

[7] Grundmann also attributes this view to Marx in the *Grundrisse* and other writings preparatory to *Capital*. In the latter, Grundmann argues, Marx abandoned this view because he had come to recognise its incompatibility with his view that capitalism is responsible for alienation.   [8] Grundmann 1991b, pp. 140–1.

[9] For an outline of such an argument see Hughes 1993, p. 38.

Grundmann allows that *institutional change* is necessary in order 'to shape the productive forces in a way which makes their detrimental effects upon the natural environment and upon human beings decrease', but he rejects the classical historical materialism which sees this institutional change primarily as a change in property relations. 'This solution', he writes, 'is fatally flawed in the light of ecological problems.'[10] His argument for this, however, is based on a misinterpretation of historical materialism, and in particular of Cohen's account of it. According to Grundmann, Cohen's account

offers only the perspective that class struggle might fight out the contradiction between productive forces and relations of production until new social relations have been established which are propitious for the productive forces. But it seems that in the case of ecological problems it is the very nature of some productive forces which causes considerable ecological damage. Hence, if we would rely on their 'autonomous' development, we would be left witnessing even more disasters.[11]

If historical materialism did, as this passage supposes, view changes in the relations of production simply as facilitating a development of the productive forces that is 'autonomous' in the sense of being an 'unmoved prime mover',[12] following its own immutable path, then Grundmann would be right to assert that it offers no prospect of resolving ecological problems. However, Grundmann is here making the same mistake as Levine and Wright;[13] he is confusing an *autonomous development* of the productive forces in the sense just explained with an *autonomous tendency* of the productive forces to develop, where the forces themselves are moulded by the prevailing social relations, but the selection of those relations ensures that in the long run the forces will develop.[14] The latter view is Cohen's, and although Cohen himself views this productive development in the narrow economic sense that Grundmann rightly views as problematic, I have pointed out that since on Cohen's model it is human interests and needs which explain the selection of production relations and thereby the development of the productive forces, and since those interests and needs include an interest in and need for a suitable environment, we can envisage the selection of structures which will facilitate the redirection of technological developments towards this end.

Grundmann's mistake is that he fails to see that a functional explanation in the manner of Cohen is compatible with social structures and human needs having an influence on the development of technology. This leads

---

[10] Grundmann 1991b, p. 222.   [11] *Ibid.*, pp. 222–3.   [12] Cf. *ibid.*, p. 162.
[13] See the discussion of Levine and Wright's misplaced criticism of Cohen in section 5.5 above.   [14] Cf. note 76, ch. 5 above.

him to draw too sharp a contrast between this explanatory dimension of Marxism, which he conceives as a form of technological determinism, and the critical dimension embodied in the ideal of productive development as an increasing capacity of humans to control their environment and satisfy their needs. Grundmann wishes to reject the former, for the reasons just outlined, while embracing the latter.[15] But in separating explanatory and critical dimensions in this way, Grundmann is ignoring the central point of Marx's critique of utopianism: that critical or normative discourse must always be integrated with an understanding of social processes. Given Marx's views on this matter it would be surprising if the two aspects of his thought could be so easily teased apart, and indeed Grundmann's attempt to do so distorts both aspects. We have already seen how he misrepresents Marx's explanatory theory by neglecting the role played by human interests and needs in shaping productive development, and conversely, in abandoning Marx's explanatory theory Grundmann leaves himself with an impoverished reading of Marx's normative theory – a reading which grasps the importance of Marx's theory of human needs, but neglects his insight into the way in which their influence is (and will continue to be) mediated by social structure.

Given this insight, which Grundmann has given us no reason to doubt, the central task of political theory is to devise structures which satisfy two conditions: firstly, that within them true human needs (including ecological needs) can exert an influence upon the ways in which technology is used and the direction of new technological developments, and secondly, that there are agents within existing social structures who will have both the motivation and the power to bring the new structures into existence. I have not, in this book, said anything about what those structures will be like, or even whether they will be recognisably socialist. However, I do think there is reason to suppose that they will involve restrictions on private property rights. I am unconvinced by suggestions that ecological problems can be solved by privatising natural resources so as to create owners with an interest in their sustainable use and a right to enforce it by charging others for using or damaging the resources. This strategy, it seems to me, is rendered unworkable by the nature of the causal relations

---

[15] Grundmann 1991b, p. 223. See also p. 234. The division between explanatory and critical dimensions of Marx's theory is also expressed in Grundmann's claim that Marx 'wavered between a social and a technological determinism'; that he 'was a technological determinist when he tried to explain historical development ("backwards"-oriented), but became a social determinist when he tried to evaluate the possibilities for a communist society ("forwards"-oriented)' (p. 200; cf. p. 166).

involved, and the difficulty in establishing the agents and actions causally responsible for particular incidents of environmental damage.

Marx, of course, is notorious for not addressing in detail the structure of a socialist society, believing the main outlines of its structure to be straightforward and the rest to be mere detail to be filled in later. Today this is widely and correctly regarded as unsatisfactory. The importance of addressing the structures of a future society is indicated by the failures and problems of socialist experiments since Marx's time, problems which need to be addressed if a viable socialist project is to be recreated. The task of devising appropriate structures is also much less straightforward than it appeared to Marx. Ecological problems make the first criterion harder to satisfy, since it is no longer possible to rely on an unexamined notion of material abundance in order to ensure the satisfaction of needs. And the second criterion is harder to satisfy because of the absence of a single homogeneous class with both the motivation and the capacity to bring new structures into existence. While the class of potential agents of such a change is broadened by the fact, emphasised by green theoreticians, that it is in everyone's objective interest to eliminate or minimise ecological problems, the opportunities for mobilising this potential are diminished by divergences within this group and by the fact that it is no longer true (if it ever was) that most of its members have nothing to lose in attempting such a change.[16] These difficulties, I would suggest, indicate an enlarged role for moral argument in bringing about new structures, and also the necessity of anticipating a stage-by-stage progress towards the favoured structure, analogous to Marx's idea of a transition from a bourgeois democratic revolution, through socialism to communism, but under different conditions. These, however, are issues to be addressed elsewhere. What I have tried to do here is to show how discussion of these issues can be situated within Marx's theory of history, and how it can be informed by Marx's analyses of society, its productive activity, and the human needs which that activity is to serve.

[16] Cf. Cohen 1995, pp. 154–8.

References

References to works by authors other than Marx and Engels are given by author and date of publication. References to Marx's and Engels's works have been given by title. This allows instant recognition of the work being cited, and avoids confusion arising from the fact that so many of Marx's and Engels's works are published in posthumous collections or editions. Marx's and Engels's works are listed alphabetically by title with date of completion given in square brackets following Marx's or Engels's name.

Abbreviations

KMSW   Karl Marx, *Selected Writings*, ed. D. McLellan (Oxford University Press, 1985).
MECW   Karl Marx and Frederick Engels, *Collected Works* (London: Lawrence and Wishart, 1975–).
MESW   Karl Marx and Frederick Engels, *Selected Works* (London: Lawrence and Wishart, 1968).

Works by Marx and Engels

Marx [1859], *A Contribution to the Critique of Political Economy* (London: Lawrence and Wishart, 1971).
Engels [1878], *Anti-Dühring* (London: Lawrence and Wishart, undated revision of 1934 edition).
Marx [1867], *Capital*, vol. I (Harmondsworth: Penguin/New Left Review, 1976).
    [1894], *Capital*, vol. III (Harmondsworth: Penguin, 1981).
    [1844], 'Critical Marginal Notes on the Article "The King of Prussia and Social Reform" by a Prussian', in *MECW*, vol. III.
    [1875], *Critique of the Gotha Programme*, in *MESW*.
Engels [1883], *Dialectics of Nature* (London: Lawrence and Wishart, 1941).
Marx [1844], *Economic and Philosophical Manuscripts of 1844* (London: Lawrence and Wishart, 1977).

Engels [1890], 'Engels to C. Schmidt in Berlin, London, October 27, 1890', in *MESW*.

[1893], 'Engels to F. Mehring in Berlin, London, July 14, 1893', in *MESW*.

[1890], 'Engels to J. Bloch in Königsberg, London, September 21[–22], 1890', in *MESW*.

[1894], 'Engels to W. Borgius in Breslau, London, January 25, 1894', in *MESW*.

Marx [1858], *Grundrisse* (Harmondsworth: Penguin/New Left Review, 1973).

Engels [1859], 'Karl Marx, A Contribution to the Critique of Political Economy', part I, in *MECW*, vol. XVI.

[1886], *Ludwig Feuerbach and the End of Classical German Philosophy*, in *MESW*.

Marx and Engels [1848], *Manifesto of the Communist Party*, in *MESW*.

Marx [1859], 'Manufactures and Commerce', *MECW*, vol. XVI.

[1851], 'Marx to Engels in Manchester. London, 7 January 1851', *MECW*, vol. XXXVIII.

[1851], 'Marx to Engels in Manchester. London, 14 August 1851', *MECW*, vol. XXXVIII.

[1862], 'Marx to Lassalle, 16 Jan 1862', in *KMSW*.

[1844], 'On James Mill', in *KMSW*.

Engels [1844], *Outlines of a Critique of Political Economy*, in *MECW*, vol. III.

Marx [1859], 'Preface' to *A Contribution to the Critique of Political Economy*, in *MESW*.

[1866], 'Results of the Immediate Process of Production', in *Capital*, vol. I.

Engels [1880], *Socialism: Utopian and Scientific*, in *MESW*.

[1883], 'Speech at the Graveside of Karl Marx', in *MESW*.

Marx [1853], 'The British Rule in India', in *Surveys from Exile* (Political Writings, vol. II), ed. D. Fernbach (Harmondsworth: Penguin, 1973).

Engels [1845], *The Condition of the Working Class in England in 1844*, in *MECW*, vol. IV.

Marx [1852], *The Eighteenth Brumaire of Louis Bonaparte*, in *MESW*.

Marx and Engels [1846], *The German Ideology*, ed. C. J. Arthur (London: Lawrence and Wishart, 1974).

Engels [1884], *The Origin of the Family, Private Property and the State*, in *MESW*.

[1876], 'The Part Played by Labour in the Transition from Ape to Man', in *MESW*.

Marx [1849], *The Poverty of Philosophy* (London: Lawrence and Wishart, no date).

[1853], 'The War Question. – British Population and Trade Returns. – Doings of Parliament', *MECW*, vol. XII.

[1861–3], *Theories of Surplus Value*, part I (London: Lawrence and Wishart, 1969); part II (London: Lawrence and Wishart, 1969); part III (London: Lawrence and Wishart, 1972).

[1845], 'Theses on Feuerbach', in *MESW*.

[1847], *Wage Labour and Capital*, in *MESW*.

[1865], *Wages, Price and Profit*, in *MESW*.

Other works cited

Archard, D. (1987), 'The Marxist Ethic of Self-Realization: Individuality and Community', in Evans (ed.).

Arthur, C. J. (1986), *Dialectics of Labour* (Oxford: Blackwell).

Attfield, R. (1991), *The Ethics of Environmental Concern*, 2nd edn (London: University of Georgia Press).

(1995), *Value, Obligation and Meta-Ethics* (Amsterdam: Rodopi).

Attfield, R. and Belsey, A., eds. (1994), *Philosophy and the Natural Environment*, Royal Institute of Philosophy supplement, 36 (Cambridge University Press).

Bahro, R. (1982), *Socialism and Survival* (London: Heretic Books).

Bakhurst, D. (1991), *Consciousness and Revolution in Soviet Philosophy* (Cambridge University Press).

Barry, B. (1965), *Political Argument* (London and Henley: Routledge and Kegan Paul).

Barry, J. (1994), 'The Limits of the Shallow and the Deep: Green Politics, Philosophy, and Praxis', *Environmental Politics* 3.

Benton, T. (1979), 'Natural Sciences and Cultural Struggle: Engels and Philosophy and the Natural Sciences', in Mepham and Ruben (eds.).

(1989), 'Marxism and Natural Limits', *New Left Review* 178.

(1991), 'The Malthusian Challenge: Ecology, Natural Limits and Human Emancipation', in Osborne (ed.).

(1992), 'Ecology, Socialism and the Mastery of Nature: A Reply to Reiner Grundmann', *New Left Review* 194.

(1993), *Natural Relations: Ecology, Animal Rights and Social Justice* (London: Verso).

Bertram, C. (1990), 'International Competition in Historical Materialism', *New Left Review* 182.

Bertram, C. and Chitty, A., eds. (1994), *Has History Ended? Fukuyama, Marx and Modernity* (Aldershot: Avebury).

Blackburn, R. J. (1990), *The Vampire of Reason* (London: Verso).

Bottomore, T., ed. (1991), *A Dictionary of Marxist Thought* (Oxford: Blackwell).

Brennan, A. (1988), *Thinking About Nature: an Investigation of Nature, Value and Ecology* (London: Routledge).

Brenner, R. (1986), 'The Social Basis of Economic Development', in Roemer (ed.).

Bugliarello, G. and Donner, D. B., eds. (1979), *The History and Philosophy of Technology* (Urbana: University of Illinois Press).

Bures, R. (1991), 'Ethical Dimensions of Human Attitudes to Nature', *Radical Philosophy* 57.

Callicott, J. B. (1989), *In Defense of the Land Ethic* (Albany: State University of New York Press).

Callinicos, A., ed. (1989), *Marxist Theory* (Oxford University Press).

Capra, F. (1983), *The Turning Point* (London: Fontana).

Carling, A. (1991), *Social Division* (London: Verso).

Casal, P. (1994), 'On Societal and Global Historical Materialism', in Bertram and Chitty (eds.).

Cohen, G. A. (1978), *Karl Marx's Theory of History: A Defence* (Oxford: Clarendon Press).

(1988), *History, Labour, and Freedom* (Oxford: Clarendon Press).

(1989), 'Reply to Elster on "Marxism, Functionalism, and Game Theory"', in Callinicos (ed.).

(1995), *Self-Ownership, Freedom and Equality* (Cambridge University Press).

Cole, H. S. D., Freeman, C., Jahoda, M. and Pavitt, K. L. R., eds. (1973), *Thinking*

About The Future: A Critique of the Limits to Growth (London: Chatto and Windus).

Collier, A. (1979), 'Materialism and Explanation in the Human Sciences', in Mepham and Ruben (eds.).

(1994), 'Value, Rationality and the Environment', *Radical Philosophy* 66.

Commoner, B. (1971), *The Closing Circle* (London: Jonathan Cape).

Daly, H. E., ed. (1980), *Economics, Ecology, Ethics* (San Francisco: Freeman).

Devall, B. and Sessions, G. (1985), *Deep Ecology* (Salt Lake City: Gibbs Smith).

Dickens, P. (1992), *Society and Nature: Towards A Green Social Theory* (London: Harvester Wheatsheaf).

Dobson, A. (1990), *Green Political Thought: An Introduction* (London: Unwin Hyman).

Dower, N., ed. (1989), *Ethics and Environmental Responsibility* (Aldershot: Avebury).

Doyal, L. and Gough, I. (1991), *A Theory of Human Need* (Basingstoke and London: Macmillan).

Eckersley, R. (1992), *Environmentalism and Political Theory* (London: UCL Press).

(1998), 'Beyond Human Racism', *Environmental Values* 7.

Ehrlich, P. R. (1968), *The Population Bomb* (New York: Ballantine).

Ehrlich, P. R. and Harriman, R. L. (1971), *How to be a Survivor: A Plan to Save Spaceship Earth* (London: Pan/Ballantine).

Ekins, P. (1991), 'Growth Without End . . . ?', *Guardian*, 18 July, p. 29.

Elliot, R., ed. (1995), *Environmental Ethics* (Oxford University Press).

Elliot, R. and Gare, A., eds. (1983), *Environmental Philosophy* (Milton Keynes: Open University Press).

Elster, J. (1985), *Making Sense of Marx* (Cambridge University Press).

(1989), 'Marxism, Functionalism, and Game Theory: The Case for Methodological Individualism', in Callinicos (ed.).

Enzensberger, H. M. (1974), 'A Critique of Political Ecology', *New Left Review* 84.

Evans, J. D. G., ed. (1987), *Moral Philosophy and Contemporary Problems*, Royal Institute of Philosophy lecture series, 22 (Cambridge University Press).

Feigl, H., Scriven, M. and Maxwell, G., eds. (1958), *Minnesota Studies in the Philosophy of Science*, vol. II (Minneapolis: University of Minnesota Press).

Fox, W. (1990), *Towards a Transpersonal Ecology* (London: Shambhala).

Freeman, C. (1977), 'Economics of Research and Development', in Spiegel-Rösing and Solla Price (eds.).

Gare, A. E. (1995), *Postmodernism and the Environmental Crisis* (London and New York: Routledge).

George, S. (1977), *How the Other Half Dies* (Harmondsworth: Penguin).

Georgescu-Roegen, N. (1980), 'The Entropy Law and the Economic Problem', in Daly (ed.).

Geras, N. (1989), 'The Controversy about Marx and Justice', in Callinicos (ed.).

(1995), *Solidarity in the Conversation of Humankind* (London: Verso).

Goldsmith, E., Allen, R., Allaby, M., Davoll, J. and Lawrence, S. (1972), *A Blueprint for Survival* (London: Tom Stacey).

Goodin, R. E. (1992), *Green Political Theory* (Cambridge: Polity Press).

Gorz, A. (1980), *Ecology as Politics* (London: Pluto Press).

Graham, K. (1992), *Karl Marx: Our Contemporary. Social Theory for a Post-Leninist World* (London: Harvester Wheatsheaf).

Green, K. (1996), 'Two Distinctions in Environmental Goodness', *Environmental Values* 5.

Grundmann, R. (1991a), 'The Ecological Challenge to Marxism', *New Left Review* 187.

(1991b), *Marxism and Ecology* (Oxford: Clarendon Press).

Hall, G. (1972a), *Ecology: Can We Survive Under Capitalism?* (New York: International Publishers).

(1972b), 'Class Aspect of the Ecological Crisis', *World Marxist Review* 8.

(1974), *The Energy Rip-Off* (New York: International Publishers).

Hardin, G. (1977), 'Living on a Lifeboat', in Hardin and Baden (eds.).

(1980), 'The Tragedy of the Commons', in Daly (ed.).

Hardin, G. and Baden, J., eds. (1977), *Managing the Commons* (San Francisco: Freeman).

Hayward, T. (1997), 'Anthropocentrism: A Misunderstood Problem', *Environmental Values* 6.

Hegel, G. W. F. (1975), *Logic* (Oxford University Press).

Heller, A. (1976), *The Theory of Need in Marx* (London: Allison and Busby).

(1985), *The Power of Shame. A Rational Perspective* (London: Routledge and Kegan Paul).

Hughes, J. (1993), 'The Red and the Green' (review of Grundmann 1991b), *Radical Philosophy* 63.

Hull, D. L. (1974), *Philosophy of Biological Science* (Englewood Cliffs, N. J.: Prentice-Hall).

Ilyenkov, E. V. (1982), *The Dialectics of the Abstract and the Concrete in Marx's Capital* (Moscow: Progress).

Irvine, S. and Ponton, A. (1988), *A Green Manifesto* (London: McDonald).

Jacobs, M. (1997), *Greening the Millennium? The New Politics of the Environment* (Oxford: Blackwell).

Jamieson, D. (1998), 'Animal Liberation is an Environmental Ethic', *Environmental Values* 7.

Jarvie, I. C. (1983), 'The Social Character of Technological Progress: Comments on Skolimowski's Paper', in Mitcham and Mackey (eds.).

Johnson, L. E. (1993), *A Morally Deep World* (Cambridge University Press).

Kemeny, J. G. and Oppenheim, P. (1956), 'On Reduction', *Philosophical Studies*, 7.

Kolakowski, L. (1969), *Marxism and Beyond* (London: Pall Mall).

(1978), *Main Currents of Marxism*, vol. I (Oxford: Clarendon Press).

Korsgaard, C. (1983), 'Two Distinctions in Goodness', *Philosophical Review* 92.

Layton, E. (1977), 'Conditions of Technological Development', in Spiegel-Rösing and Solla Price (eds.).

Lee, D. C. (1980), 'On the Marxian View of the Relationship between Man and Nature', *Environmental Ethics* 2.

(1982), 'Towards a Marxian Ecological Ethic: A Response to Two Critics', *Environmental Ethics* 4.

Lee, K. (1989), *Social Philosophy and Ecological Scarcity* (London: Routledge).

Levine, A. and Wright, E. O. (1980), 'Rationality and Class Struggle', *New Left Review* 123.

Lovelock, J. (1995), *Gaia: A New Look at Life on Earth* (Oxford University Press).

Low, N. and Gleeson, B. (1998), *Justice, Society and Nature* (London: Routledge).

Lukács, G. (1971), *History and Class Consciousness* (London: Merlin).

Lukes, S. (1985), *Marxism and Morality* (Oxford University Press).

McKibben, W. (1990), *The Death of Nature* (Harmondsworth: Penguin).

McLellan, D. and Sayers, S., eds. (1990), *Socialism and Morality* (London: Macmillan).

Malthus, T. R. (1986a), *Works*, ed. E. A. Wrigley and D. Souden, 8 vols. (London: William Pickering).

(1986b), 'An Essay on the Principle of Population', 1st edn, in Malthus (1986a), vol. I.

(1986c), 'An Essay on the Principle of Population', 6th edn, in Malthus (1986a), vols. II–III.

(1986d), 'Population', in Malthus (1986a), vol. IV.

Maslow, A. H. (1970), *Motivation and Personality*, 2nd edn (New York: Harper and Row).

Mathews, F. (1991), *The Ecological Self* (London: Routledge).

(1995), 'Value in Nature and Meaning in Life', in Elliot (ed.).

Matthews, W. H., ed. (1976a), *Outer Limits and Human Needs: Resource and Environmental Issues of Development Strategies* (Uppsala: Dag Hammarskjöld Foundation).

(1976b), 'The Concept of Outer Limits', in Matthews (ed.).

Meadows, D. H., Meadows, D. L., Randers, J. and Behrens, W. W. (1974), *The Limits to Growth* (London: Pan).

Mepham, J. and Ruben, D. H., eds. (1979), *Issues in Marxist Philosophy*, vol. II (Brighton: Harvester Press).

Mesarovic, M. and Pestel, E. (1975), *Mankind at the Turning Point* (London: Hutchinson).

Miller, R. (1981), 'Productive Forces and the Forces of Change' (review of Cohen 1978), *Philosophical Review* 90.

(1984), *Analyzing Marx* (Princeton University Press).

Mitcham, C. (1979), 'Philosophy and the History of Technology', in Bugliarello and Donner (eds.).

Mitcham, C. and Mackey, R., eds. (1983), *Philosophy and Technology* (New York: The Free Press).

Moore, S. (1975), 'Marx and Lenin as Historical Materialists', *Philosophy and Public Affairs* 4.

Naess, A. (1973), 'The Shallow and the Deep, Long-Range Ecology Movement. A Summary', *Inquiry* 16.

Naess, A. and Rothenberg, D. (1989), *Ecology, Community and Lifestyle: Outline of an Ecosophy* (Cambridge University Press).

Nagel, E. (1961), *The Structure of Science* (London: Routledge and Kegan Paul).

(1979), *Teleology Revisited and Other Essays in the Philosophy of Science* (New York: Columbia University Press).

Naletov, I. (1984), *Alternatives to Positivism* (Moscow: Progress).

Norman, R. (1996), 'Interfering with Nature', *Journal of Applied Philosophy* 3.

Odum, E. P. (1975), *Ecology* (London: Holt, Rinehart and Winston).

O'Neill, J. (1993), *Ecology, Policy and Politics: Human Wellbeing and the Natural World* (London: Routledge).

O'Neill, O. (1986), *Faces of Hunger: An Essay on Poverty, Justice and Development* (London: Allen and Unwin).

Oppenheim, P. and Putnam, H. (1958), 'Unity of Science as a Working Hypothesis', in Feigl, Scriven and Maxwell (eds.).

O'Riordan, T. (1976), *Environmentalism* (London: Pion).

Osborne, P., ed. (1991), *Socialism and the Limits of Liberalism* (London: Verso).

Parsons, H. L., ed. (1977), *Marx and Engels on Ecology* (Westport, Conn.: Greenwood Press).

Passmore, J. (1974), *Man's Responsibility for Nature: Ecological Problems and Western Traditions* (London: Duckworth).

Pearce, D., Markandya, A. and Barbier, E. B. (1990), *Blueprint for a Green Economy* (London: Earthscan).

Pepper, D. (1993), 'Anthropocentrism, Humanism and Eco-Socialism: A Blueprint for the Survival of Green Politics', *Environmental Politics* 2.

Plamenatz, J. (1963), *German Marxism and Russian Communism* (London: Longman).

Plant, R., Lesser, H. and Taylor-Gooby, P. (1980), *Political Philosophy and Social Welfare* (London: Routledge and Kegan Paul).

Plekhanov, G. (1937), *Fundamental Problems of Marxism* (London: Lawrence and Wishart).

Porritt, J. (1985), *Seeing Green: The Politics of Ecology Explained* (Oxford: Blackwell).

(1991), 'Global Warning', *New Statesman and Society*, 10 May, p. 15.

Prinz, A. M. (1969), 'Background and Ulterior Motive of Marx's "Preface" of 1859', *Journal of the History of Ideas* 30.

Raz, J. (1984), 'Right-Based Moralities', in Waldron (ed.).

(1986), *The Morality of Freedom* (Oxford University Press).

Redclift, M. (1989), 'Turning Nightmares into Dreams: The Green Movement in Eastern Europe', *The Ecologist* 19.

Regan, T. (1988), *The Case for Animal Rights* (London: Routledge).

Rifkin, J. and Howard, T. (1980), *Entropy: A New World View* (New York: Viking Press).

Roberts, M. (1996), *Analytical Marxism: A Critique* (London: Verso).

Roemer, J. E. (1982), 'Methodological Individualism and Deductive Marxism', *Theory and Society* 11.

Roemer, J. E., ed. (1986), *Analytical Marxism* (Cambridge University Press).

Rolston, H. (1994), 'Value in Nature and the Nature of Value', in Attfield and Belsey (eds.).

Routley, V. (1981), 'On Karl Marx as an Environmental Hero', *Environmental Ethics* 3.

Ruben, D. H. (1979), *Marxism and Materialism* (Brighton: Harvester Press).

Sayers, S. (1984), 'Marxism and the Dialectical Method: A Critique of G. A. Cohen', in Sayers and Osborne (eds.).

(1991), 'F. H. Bradley and the Concept of Relative Truth', *Radical Philosophy* 59.

Sayers, S. and Osborne, P., eds. (1990), *Socialism, Feminism and Philosophy: A Radical Philosophy Reader* (London: Routledge).

Schaffner, K. F. (1967), 'Approaches to Reduction', *Philosophy of Science* 34.

Schmidt, A. (1971), *The Concept of Nature in Marx* (London: New Left Books).

Searle, J. R. (1995), *The Construction of Social Reality* (London: Allen Lane).

Sen, A. (1982), *Poverty and Famines* (Oxford: Clarendon Press).

Shaw, W. H. (1978), *Marx's Theory of History* (Stanford University Press).

(1991), 'Historical Materialism', in Bottomore (ed.).

Singer, P. (1993), *Practical Ethics*, 2nd edn (Cambridge University Press).

Skolimowski, H. (1981), *Eco-Philosophy: Designing New Tactics for Living* (Boston: Marion Boyars).

(1983), 'The Structure of Thinking in Technology', in Mitcham and Mackey (eds.).

Sober, E. (1995), 'Philosophical Problems for Environmentalism', in Elliot (ed.).

Spiegel-Rösing, I. and Solla Price, D. de, eds. (1977), *Science, Technology and Society: A Cross-Disciplinary Perspective* (London: Sage).

Spretnak, C. and Capra, F. (1986). *Green Politics* (London: Paladin).

Springborg, P. (1981), *The Problem of Human Needs and the Critique of Civilization* (London: Allen and Unwin).

Sylvan, R. and Bennett, D. (1994), *The Greening of Ethics* (Cambridge: The White Horse Press).

Taylor, P. W. (1986), *Respect for Nature: A Theory of Environmental Ethics* (Princeton University Press).

Timpanaro, S. (1975), *On Materialism* (London: New Left Books).

Tolman, C. (1981), 'Karl Marx, Alienation and the Mastery of Nature', *Environmental Ethics* 3.

Van Parijs, P. (1982), 'Functionalist Marxism Rehabilitated: A Comment on Elster', *Theory and Society* 11.

(1993), *Marxism Recycled* (Cambridge University Press).

Waldron, J. (1984), *Theories of Rights* (Oxford University Press).

Walker, K. J. (1979), 'Ecological Limits and Marxist Thought', *Politics* 14.

Watson, R. (1983), 'A Critique of Non-Anthropocentric Biocentrism', *Environmental Ethics* 5.

Wiggins, D. (1987), *Needs, Values, Truth* (Oxford: Blackwell).

Wilde, L. (1998), *Ethical Marxism and its Radical Critics* (Basingstoke: Macmillan).

World Commission on Environment and Development (1987), *Our Common Future* (Oxford University Press).

Wright, L. (1973), 'Functions', *Philosophical Review* 82.

Index